建筑立场系列丛书 No.25

# 在城市中转换
# Variation and Transition

中文版

韩国C3出版公社 | 编

# 城市设计
## 在城市中转换
026 建筑面积和城市影响力 _ Silvio Carta

### 都市生活体验的场景
028 都市生活体验的场景 _ Paula Melâneo
034 飞人屋 _ Louis Paillard Architect & Urbanist
046 日本潮牌办公大楼 _ General Design
056 隐藏的住宅 _ Teatum + Teatum
060 阶梯住宅 _ Jaques Moussafir Architectes
068 狄维娜护理之家14 _ Estudio de Arquitectura Javier Terrados
074 巴洛克庭院公寓 _ OFIS Arhitekti
082 萨莫拉办公楼 _ Alberto Campo Baeza

### 面朝街道，心系大海
098 面朝街道，心系大海 _ Diego Terna
104 27D大楼 _ Kraus Schönberg Architects
110 墨西哥城西班牙文化中心 _ JSª Arquitectura + Arquitectura 911SC
118 梦想商业中心酒店 _ Handel Architects
126 巴黎篮子学生公寓 _ OFIS Arhitekti

### 层次感活跃了城市表面
136 重新思考地面层 _ Simone Corda
140 塞万提斯大剧院 _ Ensamble Studio
150 Teruel-Zilla文化休闲中心 _ Mi5 Arquitectos + PKMN Arquitectura
164 Georges-Emile-Lapalme文化中心
_ Menkès Shooner Dagenais LeTourneux Architectes + Provencher Roy+Associés Architectes
170 南阿尔伯塔理工学院停车场 _ Bing Thom Architects

### 资讯
004 森林 _ Atelier WHY
008 同一屋檐下 _ Kengo Kuma & Associates + Holzer Kobler Architekturen
010 阿富汗国家博物馆 _ AV62 Arquitectos
014 生命之家 _ Tomas Ghisellini Architects
018 St. Horto花园 _ OFL Architecture
020 新加坡2012建筑节零浪费展馆 _ WOW Architects

180 建筑师索引

C3 建筑立场系列丛书 No.25

# UrbanHow
## Variation and Transition

026 *Building Extent and Urban Influence _ Silvio Carta*

### Scenario for Urban Daily Life Experience
028 *Scenario for Urban Daily Life Experience _ Paula Melâneo*
034 The Trapeze House _ Louis Paillard Architect & Urbanist
046 Neighborhood Office Building _ General Design
056 Hidden House _ Teatum + Teatum
060 Step House _ Jaques Moussafir Architectes
068 Divina Enfermera 14 _ Estudio de Arquitectura Javier Terrados
074 Baroque Court Apartments _ OFIS Arhitekti
082 Zamora Offices _ Alberto Campo Baeza

### Facing the Street, I am (not) a Duck
098 *Facing the Street, I am (not) a Duck _ Diego Terna*
104 House 27D _ Kraus Schönberg Architects
110 Spain's Cultural Center in Mexico _ JSª Arquitectura + Arquitectura 911SC
118 Dream Downtown Hotel _ Handel Architects
126 Basket Apartments in Paris _ OFIS Arhitekti

### Layering Active Urban Surface
136 *Re-thinking the Ground Level(s) _ Simone Corda*
140 Cervantes Theater _ Ensamble Studio
150 Teruel-Zilla _ Mi5 Arquitectos + PKMN Arquitectura
164 Georges-Emile-Lapalme Cultural Center
_ Menkès Shooner Dagenais LeTourneux Architectes + Provencher Roy+Associés Architectes
170 SAIT Polytechnic Parkade _ Bing Thom Architects

### News
004 The Forest _ Atelier WHY
008 Under One Roof _ Kengo Kuma & Associates + Holzer Kobler Architekturen
010 Afghanistan National Museum _ AV62 Arquitectos
014 Home for Life _ Tomas Ghisellini Architects
018 St. Horto _ OFL Architecture
020 Archifest Zero Waste Pavilion _ WOW Architects

180 Index

# 森林 _Atelier WHY

美国建筑师协会底特律分会的城市优先事物委员会（AIA-UPC）最近宣布了底特律河滨设计竞赛的获奖者名单，韩国设计师HyunTek Yoon和SooBum You（WHY工作室）的作品荣获一等奖。充分合理利用每一寸空间是城市发展规划的美德，也成为底特律的象征。城市的快速发展就是努力充分合理利用每一寸空间的结果。项目地点位于整个放射性街道规划的节点上，目前这里充斥着树木和小山丘，杂乱无章。森林给城市带来了无穷的遐想，这个森林公园拥有自然元素以及像树木和小山丘这些可以利用的公共空间。空间尺度变化和各种各样的活动将给人们带来丰富多样的体验。

### 林墙环绕

整个森林公园的边缘将由高大的树木紧密环绕，构成与城市文脉强烈的对比效果。稠密的树木给整个森林公园蒙上一层神秘的面纱，将使人们无法看到树木另一边的风景，激起人们的好奇心和诗情画意般的想象力，使他们感到仿佛置身于仙境或未知的世界。从根本上说，很难用像城市街道上一些标示、入口或人行道之类的任何做法来控制游客的行进路线。人们会慢慢地走在蜿蜒穿过森林的小步道上，踩着松软的泥土，悠闲而漫无边际。郁郁葱葱的森林在规模上会让通行于公园边上宽敞的主要街道或杰佛逊大道上的人们感受到强大的视觉冲击力。

### 林中音乐会

你曾想象过在大树下聆听一场小型演唱会吗？柔和的阳光洒满小径，空气中回荡着从三角钢琴中溢出的甜蜜而经典的旋律，音乐家和观众共同感受到令人无比愉悦的视听盛宴。音乐会场经过精心设计，与森林的自然环境和谐相融。首先，音乐会座椅被设计成层层波浪般的树根，就好像周围大树的树根沿着小路自然舒展延伸，细致入微的纹理和造型营造出梦幻中或童话故事中才有的古老森林的氛围。另外，一座窄窄的木结构小桥架于小路之上，人们可以在上面休息放松和观赏演出。

### 野口雕塑

艺术家野口勇的两个雕塑将被用于装饰历史和文化空间。喷泉所使用的材料和设计的形状都非常独特，低于林中小路约2.7m，位于一个下沉花园和小型圆形剧场之内，四周由一圈圆形结构包围。随风摆动的窗帘将营造一种诗意的氛围，让游客联想到野口勇喷泉雕塑的水帘。同时，地下室有餐馆和零售商店等商业设施。

### 小山丘观景台和观景台下面

小山丘将被用作室外剧场、公共集市和举行文化活动之所。观景台提供了一处如雨

篷一般的半封闭空间，可以作为门厅和遮风挡雨的地方。特别是在雨雪天气里，这里就可以替代森林中和小山丘上的室外空间了。此外，这里也将成为市民和游客最活跃的晚上活动空间。自动扶梯直接通到小山丘屋顶的观景台，给游客一种前所未有的非凡体验。

**森林和河滨**

小山丘观景台将森林与河滨连为一体，从森林到大湖边，让人们体验到时空的变换。小山丘观景台是整个景观的一部分，但又是河滨的标志性建筑。在这里，人们不仅有线性体验，还将有垂直和水平的空间体验，但人们体验的空间美感将取决于身处何处和所做何事。这个观景台将被用作森林公园的游客中心、小礼堂和公园后勤保障处。

## The Forest

American Institute of Architects Detroit's Urban Priorities Committee (AIA-UPC) has recently announced the winners of the Detroit by Design 2012 Detroit Riverfront Competition and Korean architects Hyun-Tek Yoon and SooBum You(Atelier WHY) have received the first prize.

The act of "filling" is the virtue of urban development and those became the symbol of Detroit. A rapid pace of growth is the result of such efforts to fill space. The site is located at the node of the radial street plan. Currently, the site is filled with voids, such as trees and the knoll. The forest creates rich stories with the city. The site has natural elements and a public space that is represented by the trees and the knoll. People will have diverse experiences based on the spatial scale variation and variety of activities.

### Edge-Forest Wall

On the edge of the park, the area will be filled with tall trees to create a deep contrast with the urban context. People will not be able to notice what is going to be unfolded to them beyond the forest due to the density of trees. This will arouse curiosity and poetic imagination in people as if they are in Wonderland or in unknown world. Radically, it will be difficult to establish any attempts to control the movement

of visitors such as signs and entrances or pedestrian roads on the street. People will meander slowly on small trails through the forest, which will be covered with soil and dirt without any determined direction. The density of the forest provides people who approach the park along the wide main streets or Jefferson Avenue with a strong reversal of perception in terms of scale.

### Concert in the Forest

Have you ever imagined a small concert under big trees? Soft sunlight fills the trails and a sweet classical melody from a grand piano resonates with the space to stimulate the emotional sensitivity of both musicians and spectators. The concert facilities are sophisticatedly designed to be in harmony with the forest's contextual nature. First, concert seats are designed as if they are the wavy roots of big trees and are distributed naturally along the trails. There concrete textured shapes create an atmosphere of an old forest that existed in fantasy lands or in fairy tales. Also, a thin bridge composed of a wood structure will string itself over a trail on which people can relax and enjoy the performances.

### Noguchi's Sculpture

Isamu Noguchi's two sculptures will be used as historical and cultural spaces. The fountain, with its unique shape and material, will be moved down nine feet under the trail level and will be located within a sunken garden and a small amphitheater surrounded by another circle structure. The movement of the curtain will create a poetic atmosphere and remind visitors of the falling water of Noguchi's sculpture. Also, the basement has commercial facilities such as restaurants and retail shops.

### The Knoll and Under the Knoll

The knoll will be used as exterior theater, public market, and cultural activities. The knoll provides a semi-exterior space like a canopy. This space will be used as a shelter and a foyer. In particular, in rainy and snowy weather, this space can be an alternative area of exterior space in the forest and in the knoll. In addition, it will be the most active space for night venues. The escalator is connected directly with the roof of the knoll. This spatial transition gives a radical and extraordinary experience.

### Forest and Riverfront

The knoll will integrate the forest and Riverfront. People will experience a spatial transition from the forest to the Great Lake. The knoll is part of the landscape, but at the waterfront, it is read as an iconic building itself. People will have vertical and horizontal experiences, as well as linear experiences. The image of space that people experience will depend on where they are located and what they are doing. This building will be used as a visitor center, small auditorium, and back of house for the Forest.

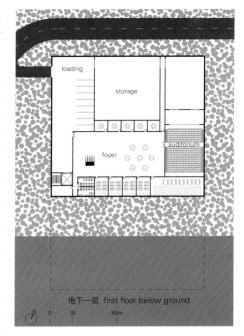

项目名称：The Forest
地点：Hart Plaza, Detroit, Michigan, USA
建筑师：HyunTek Yoon, SooBum You (Atelier WHY)
用途：multipurpose public space
用地面积：65,500m²
建筑面积：6,000m²
总建筑面积：8,050m²

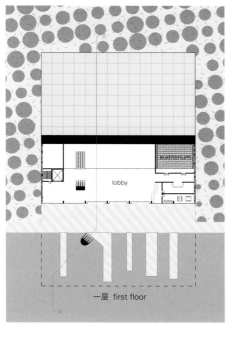
一层 first floor

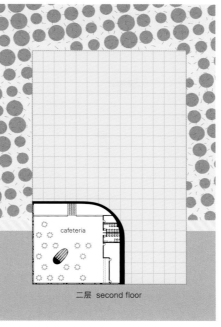
二层 second floor

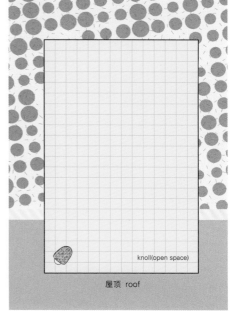
屋顶 roof knoll(open space)

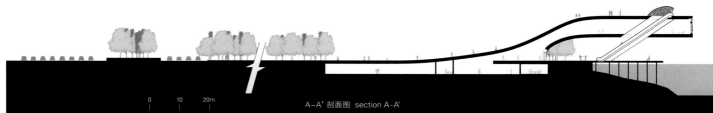
0 10 20m  A-A' 剖面图 section A-A'

# 城市干预 URBAN INTERVENTION

## 同一屋檐下 _Kengo Kuma & Associates + Holzer Kobler Architekturen

日本建筑公司Kengo Kuma & Associates连同苏黎世Holzer Kobler建筑师事务所一起赢得了2012年所举办的为洛桑联邦理工大学(Espaces et pavillons sur la place Cosandey)设计一个科桑德广场(Cosandey Square)的建筑设计大赛。获奖项目"同一屋檐下"将一个实验性艺术与科学的空间和一个展示馆设置在与蒙特勒爵士音乐实验室同一个长条状石质屋檐下。

该项目位于一大片草地上，是洛桑联邦理工大学校园中央的一块空地。该项目将校园的北面区域（游憩广场、校园社交中心以及有轨电车站）和南面的学生公寓分隔开来。同样，它还将结构密集的西面区域与围绕着各种各样的学习中心迅速发展起来的东面区域分隔开来。由于项目基地的面积广阔，因此展馆的位置和结构可有多种选择。建筑师们决定把三个要修建的展馆集于一体，建成一个非常狭长的建筑，把原来毫无功能可言的中央空地转变成一个校园连接枢纽：

——整个屋顶全长270m，北接游憩广场，南连学生公寓，学生一天几次穿行于二者之间，可以利用该屋顶遮风挡雨。

——屋顶下的门廊位于两座体量之间，对应西面的主道，通向主要的公共停车区域，以及东面目前正在施工中的林荫大道。因此，门廊使该建筑具有了通透性，它吸引并连接着校园两侧的区域。

通过改造，"同一屋檐下"变成了学生、教授和游客每天愉快经过之地，可以欣赏这儿举行的新鲜丰富的文化活动。整个区域将成为校园的中央枢纽，丰富洛桑联邦理工大学的社交性和文化性。

该展馆将于2013年春季由总承包商Marti S.A.负责建造，并计划于2014年秋季对外开放。

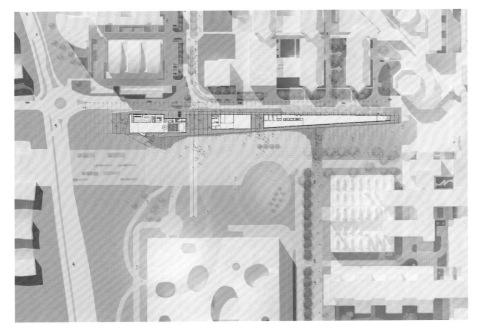

项目名称：Under One Roof
地点：Lausanne, Switzerland
建筑师：Kengo Kuma & Associates, Holzer Kobler Architekturen
项目规划：reception venue, exhibition space, music hall, restaurant
总建筑面积：3,500m²

## Under One Roof

The Japanese architectural firm Kengo Kuma & Associates, together with Holzer Kobler Architekturen from Zurich, won the architectural design competition launched in 2012 to develop Cosandey Square at EPFL (Espaces et pavillons sur la place Cosandey). The winning project, "Under One Roof," will unite an experimental Art & Sciences space and a demonstration pavilion under a single, long stone roof at the Montreux Jazz Lab.

The project site is a vast lawn, a void in the middle of the EPFL campus. It disconnects the North side of the campus (where the Esplanade plaza, social heart of the campus, and the tram station are) from the students' residential area on the South. Also it separates the dense west part of the campus with the currently evolving East side that is articulated around the impressive presence of the Learning Center.

This vast project site enables the pavilions in many possible locations and configurations. The architects decided to gather the three required pavilions into one very long and thin building that would transform the site from being a dysfunctional void into a campus connection hub:

- The 270 m long roof will shelter and go along with the students walking flow from the north Esplanade plaza down South to their residences several times a day.
- The porches provided between the volumes under the roof will be aligned to the main street coming from the West side, leading to the main public parking areas, and to the new tree avenue coming from the East, currently under construction. Therefore, the porches will provide permeability through the building attracting and connecting these both sides of the campus. By transforming the site into a place where students, professors and visitors will pleasantly pass by everyday enjoying the new cultural activities that will take place under this roof. This whole area will become a central hub within the campus and will bring a more social and cultural dimension to the EPFL.

The construction of the pavilion will be carried out by the general contractor Marti S.A. and will begin in the spring of 2013, and its doors are scheduled to open in the autumn of 2014.

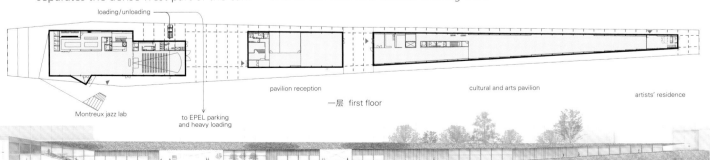

## 阿富汗国家博物馆 _AV62 Arquitectos

阿富汗信息和文化产业部与美国驻喀布尔大使馆已经评选出AV62建筑师事务所作为为阿富汗国家博物馆国际建筑设计竞赛的获奖者。

建筑的规模和扩建以及变革的速度都在迅猛增长，这些都是源于科学和技术的进步。尽管这种现状充满着挑战，且在其他人无法选择或理解的情况下，建筑师最终设计出晦涩难懂、与世孤立和背离主题的作品。

在某种程度上，这使他们认为，因为不同的原因，在几乎完全不一样的背景下，他们在世界上所面临的挑战与阿富汗正面临的挑战是一样的：在构建他们的生活中，无法获得一种高效且有慈悲情怀的知识。

19世纪文化设施的概念是为公民提供一个稳定而封闭的空间来获得广泛认可的学术知识，现在是时候创建另一种空间形式了。这一空间将有助于人们真正聚在一起分享他们所获得的各个层面的知识，包括智力方面的、经验的、审美的、情感的。在这里，人们集体为了全人类的福祉，进行富有象征意义的图像和信息的创作和传播，从中获得理解他们想要改变的世界的工具。

博物馆的主要目的是培养能够充分关注所居住环境并积极参与其中的人。

要提高语言代码的读写能力，构建我们共同的未来，就有必要帮助人们理解语言代码从过去发展到现在的过程。建筑总是重建，每个社会都需要知道重新开始需要的东西。就阿富汗来说，这一点更加富有戏剧性，更加紧急。

该项目的首要目标就是提高建筑物的开放性与互动性。建筑师认为，这个建筑不应被视作一个空间密集的、封闭的、自我指涉和先验代表的建筑。他们必须想出一个策略，如何使人们经过这个空间就能将它认出来。这一想法也让建筑师们想到这样一个空间，在

这一空间里，建筑不会从一开始就只是一个形式。灵活的空间能够随着时间的流逝而进行改变，建筑师们不停地思考，并以像科尔多瓦清真寺这样的一个地方作为参照。

围墙也许看起来不受欢迎，但为确保安全又必不可少，是确定清晰边界的关键元素，使建筑师们可以创建一个完整的室内世界，可以把四合院或清真寺的传统结合起来。四周的围墙可以把自然和生命保护在这一方寸空间内，远离充满敌意的环境。

围墙之内，是像清真寺、集市和四合院一样灵活的拓扑结构，因为构建的是一个个结构模块，因此可以对原来没有预料到的要求做出快速响应。

屋顶的作用是包容和保护不能裸露于外界的东西，相当于花园里的树木，博物馆的屋顶设计从空中看确实如此。镶有瓷砖的屋顶替代了自然，它用几何图形表示着大自然，阐释着大自然。

### Afghanistan National Museum

The Afghan Ministry of Information and Culture and the US Embassy in Kabul has

chosen AV62 Arquitectos as the winner of the International Architectural Competition for the National Museum of Afghanistan.

The size and proliferation of data, the speed of the changes resulting from advances in science and technology is growing exponentially. This situation, although challenging, led the architect to the opacity, the isolation and the irrelevance, by our inability to select and therefore to understand.

Somehow this leads them to think that, for different reasons, and in almost opposite contexts, the challenges they face in the world are the same ones Afghanistan is facing: inability to access to a knowledge capable of accompany with efficiency and compassion in the construction of their lives.

Exceeded the nineteenth-century concept of cultural facilities that make available to the citizen an agreed academic knowledge, stable and closed, it's time to make spaces for meeting and debate where is conducive the real meeting of people and sharing of knowledge at various levels including intellectual, experiential, aesthetic, emotional factors. Places for collective work for creation and dissemination of symbolic images and messages that allow us to begin to draw tools for the under-

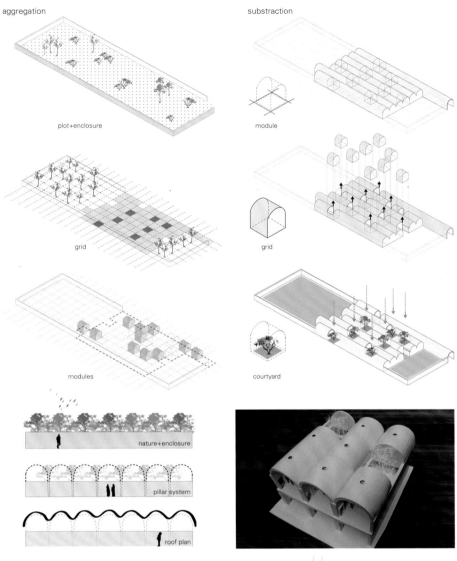

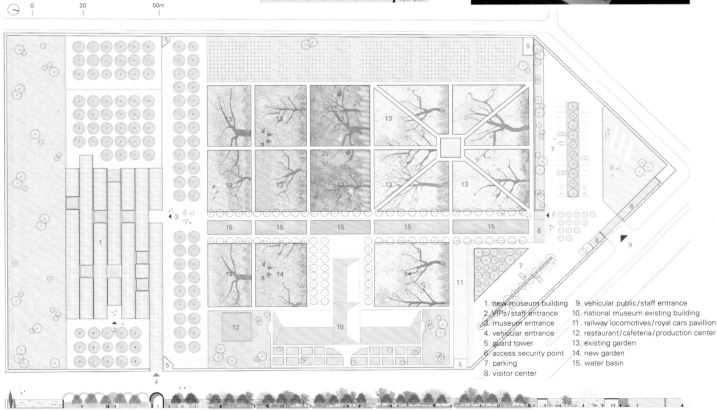

1. new museum building
2. VIPs/staff entrance
3. museum entrance
4. vehicular entrance
5. guard tower
6. access security point
7. parking
8. visitor center
9. vehicular public/staff entrance
10. national museum existing building
11. railway locomotives/royal cars pavillion
12. restaurant/cafeteria/production center
13. existing garden
14. new garden
15. water basin

standing of a world they want to transform for the benefit of all.

The main purpose of a museum will be the training of persons who could be active and conscious part of their environment.

To facilitate literacy in the language codes that are to build our shared future, it is necessary to facilitate the understanding of what shapes them into the present from the past. The construction is always rebuilding, and every society needs to know what is available to start over. In the case of Afghanistan it is even more dramatic and urgent.

The first and most important objective of this project is to increase the openness and dialogue of the building. The architect thinks that this building should not be thought of as a tight, closed, self-referential and a priori representative object. They must think of a strategy of occupying space that allows them to recognize the space as you go by. This also leads them to think of a space in which architecture doesn't aspired to be a form from the beginning. Flexible spaces capable to be adapted to the passing of time and we can not stop thinking about a place like the Mosque of Cordoba as a referent.

The enclosing wall might seem undesirable imperative linked to security, but is to them a key element for defining a clear boundary and allows them to create a complete interior universe. This allows them connecting with the tradition of the courtyard house or the mosque. A perimeter that allows them confining nature and life to preserve and protect it from a hostile environment.

Within this wall the flexible topologies like the ones of mosques, markets and courtyard houses, which can respond quickly because of the establishment of a structural module to the unexpected requirements.

The cover is the element that contains and protects the program that is not developed in the open air. The cover is equivalent to the trees of the garden and so is seen from the sky. A covering of ceramic tile, which interprets and geometrizes the nature it is replacing.

项目名称：Museo Nacional de Afghanistán, Kabul
地点：Kabul, Afghanistán
建筑师：AV62 Arquitectos
项目团队：Stefano Carnelli, Blanca Pujals, Itziar Imaz, Samantha Sgueglia, Nuno Lopes
效果图：Luis de la Fuente
机械与电气工程师：BIS arquitectes
用地面积：15,000m²
建筑面积：9,300m²

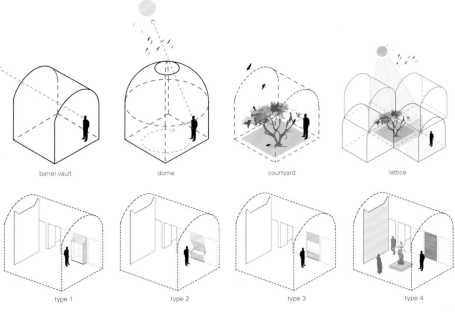

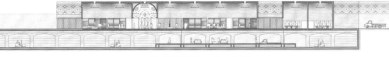

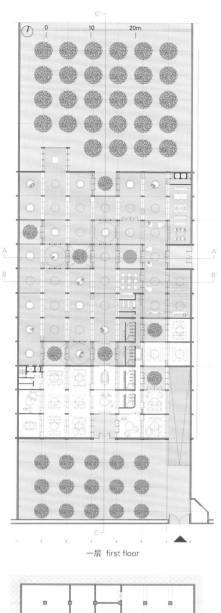

一层 first floor

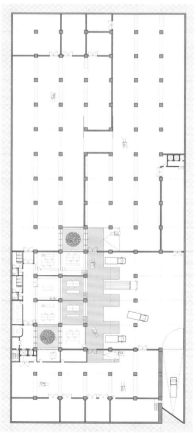

地下一层 first floor below ground

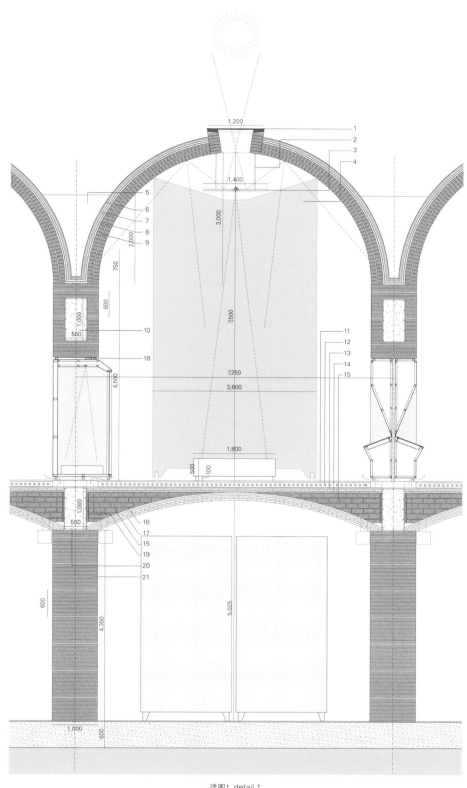

详图1 detail 1

1. support and insulation structure for skylight (diameter 1,200mm), "climalit" 4+4/7/5+5
2. wires to support the circular structure
3. circular structure which allows indirect natural illumination, as well as adjustable artificial light
4. mobile structure for division and organization of the exhibition space
5. sinus stuffed vaults, brick
6. coating in ceramic tiled
7. thermical natural isolation
8. layers of compression reinforced concrete
9. two layers of bricks on edge with mortar
10. mixed pillars of brick and reinforced concrete, the ceramic involves and serves as the concrete formwork.
11. continuous terrazzo/afghan limestone
12. radiating floor: a system of buried flexible pipes provides either heating of cooling for thermal comfort of the users. It is a high efficiency and high comfort system, can uniform temperature distributing thanks to radiating floor. The whole floor acts as radiation element.
13. laters of compression reinforced concrete mesh
14. two layers of ceramic tongued
15. solid brick wall
16. layers of compression reinforced concrete
17. timbrel vault: three layers of tiled vault with mortar joints
18. canal for facilities. sprinklers/security cameras
19. ceramic formwork
20. reinforced concrete main beam
21. mixed pillars of brick and reinforced concrete, the ceramic involves and serves as the concrete formwork. brick modules:
   - main structural ceramics: 24x12x6cm
   - timbrel vault ceramics: 24x12x3cm

## 生命之家 _Tomas Ghisellini Architects

这是一个全新的城市殡仪馆和社会基础设施综合体的设计方案,由Tomas Ghisellini建筑师事务所设计,赢得了生命之家(Domus Vitae)的设计一等奖。该项目旨在重新解读众多城市问题中的一个,也许这个问题更密切地植根于人们心里对他们的城市所持有的印象。这个项目将重新建造属于埃斯特家族的漂亮的Delizie住宅(一栋拥有大大花园的乡间别墅)。

代表建筑物边界的墙经过凿刻,可以说是透明的,无论是步行或骑自行车,路人都会禁不住好奇地从外面窥探一下这一大型绿色空间,这成为他们经历的一部分。历史上著名的费拉拉花园,原来是围墙包围的一个与世隔绝、排斥外界的地方,现在发展成为人们可以在此相聚交流的社交空间,或者说是一块流畅的公共城市地毯。

新建筑是一种连续而多孔的建筑形式:阳台、门廊、天井、露台、悬臂结构和悬挑体量等能够捕捉或放大自然光线,形成大气质量良好的空间,大气质量是其关键性的附加效果。

紧挨着南部边缘的一座现有建筑,嵌入了一条线性技术服务带,包括对建筑综合体来说所有必需的技术设备和服务功能(寄存处、储藏室、技术箱、卫生间、机械设备间、垂直交通和服务入口)。主入口位于新旧建筑之间,沐浴在上方照射下来的自然光之中,为员工专设进出通道,让人联想到具有浓郁历史风情的小巷。

现有的南侧建筑能够显示出其北立面的内部布局,其中设有接待室、观察室、尸体分析储存室以及员工行政、管理和保障室。工作人员的娱乐设施位于建筑东侧,靠近一个小型的公共空间,可以从外部直接进入,其中包含一个咖啡店和一个小吃店,到此的哀悼者和偶尔经过的访客可以在此小憩。此外,居民晚上可以在这里聚在一起聊天,喝杯咖啡,或者只是默默地坐在花园里放松身心。

原有的圆形旧矿坑现在被建成一个大型的矿物室外天井,成为新建筑的重心,并使这里成为最引人注目的室外公共聚会区域,哀悼者可以在此相聚。接待处在一楼,设有接受和移交遗骸的地方,还有用于准备遗骸告别仪式的房间。环绕双层高的门厅,面对着天井和东面历史上著名的城池防御墙,楼梯间和升降机间从地下室直上直下一直通到最高处,没有游客,也不会接触到员工。

守夜者的空间在这个项目中被设计成充满阳光的房间,摆脱了西方传统中守夜者空间的全封闭式特点,同时又确保人们在自省时不会受到打扰:一整面墙全是玻璃,从里面可以看到美丽的向天空开放的带有空中花园的双层秘密天井,有花有树。这五个情感空间环境都让哀悼者感到亲切,给哀悼者痛苦的经历留下些许"安慰"。

每个秘密天井里都有一位当代艺术家的作品;太平间建有情感分享的地方,多亏有了艺术语言,人们生活的空间才变得具有诗意。

一条室外空中"通道",穿过矿物天井,通到没有其他方式可以到达的一个神秘的户外瞭望台。瞭望台面对花园,可以看到文艺复兴时期的城墙的轮廓。这一特别的冥想空间专为个人独处和沉思默祷设计。悬浮在透明的建筑立面正对面,面对冉冉升起的太阳,这一悬浮建筑体环绕着庭院,漂浮在空中,拥抱着访客。

在原有的古老圆形矿坑处种植一颗神圣的树再合适不过,在所有文化和宗教信仰中,树都是生命和重生的象征,所以在此人们举行死亡仪式不是把死亡看作生命的终结,而仅仅是生命形式的转换。因此,我们把它叫作Domus Vitae——生命之家。

项目名称: Domus Vitae
地点: Ferrara, Italy
建筑师: Tomas Ghisellini Architects
合作者: Michele Marchi, Alice Marzola
结构: Beatrice Bergamini
设备与消防安全: Nicola Gallini
可持续性及LEED评估: Violeta Archer
甲方: Municipality of Ferrara
用地面积: 9,730m²
建筑面积: 1,560m²
总建筑面积: new+recovery – 1,290m², underground 1,590m²
项目规划: city morgue, social facilities complex
竣工时间: 2016

## Home for Life

Designed by Tomas Ghisellini Architects, the first prize winning proposal for the Domus Vitae, a new city morgue and social facilities complex, is aimed at reinterpreting one of the urban issues perhaps more intimately rooted into the mental image that people keep of their city. This project is the regeneration of wonderful Delizie (a country house with huge gardens) of the Este Family.

The border wall is carved and made literally transparent; passersby, on foot or by bicycle, intrigued by the opportunity to spy on the large green space from the outside, becoming part of the experience. The historic Ferrara walled garden, from a territory of separation and exclusion, evolves into a social space to meet, a collective and fluid urban carpet.

The new architectural presence is a continuous but porous body: balconies, porches, patios, terraces, overhangs and suspended volumes capture, tame or magnify natural light, creating spaces for which the atmospheric quality is supposed to be a decisive added value.

Flanked to one of the existing buildings along the southern edge, a plug-linear technology spine incorporates all the technical equipments and service functions necessary to the complex (deposits, storages, technical boxes, toilets, plant rooms, vertical connections, service entrances) and the approach-gap conserved between old and new, illuminated by natural light raining from above, distributes the spaces reserved for the sole employees arousing the perceptive suggestion of a historic alley.

The existing southern building shows to the these inner distributions its north elevation. It hosts

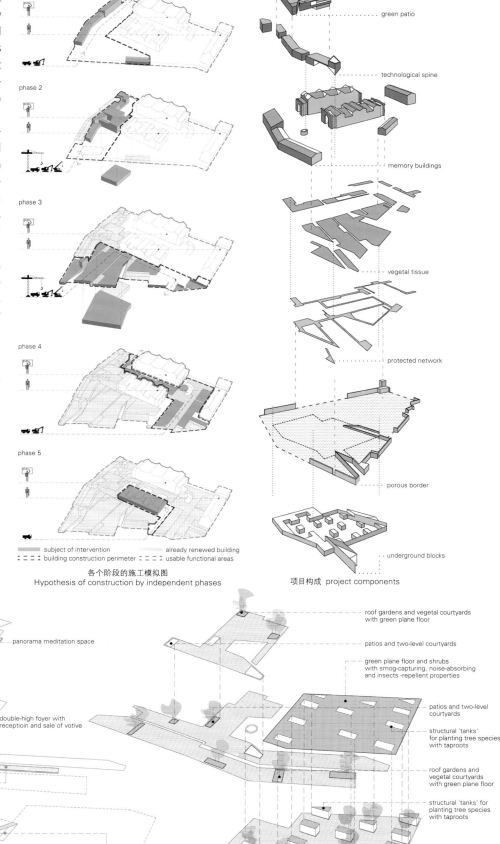

functions of acceptance, observation, analysis and storage of corpses, as well as the administrative, management and support to the personnel whose recreational facilities are strategically positioned on the east, close to a small public space, accessible from the outside, reserved to a coffee and snack bar. This cafeteria will also refresh mourners and occasional visitors to the citadel. Here, moreover, residents will gather in the evenings to chat, have a coffee, or just relax silently on the gardens.

A large mineral outdoor patio embraces the old circular pit making it become the new composition's center of gravity, and drawing here the most significant common meeting area for mourners on the outside. The ground floor hosts the reception and sets up the places for acceptance and movement, as well as ceremonial rooms used in the preparation of remains. Around the double-height foyer, facing the patio and the historic city defensive walls to the east, stairs and lifters blocks allow vertical displacements from the basement straight up to the highest nobel level without visitors and staff never come into contact.

Spaces for the wake, away from the hermetic character of the Western tradition, yet perfectly protected from any introspection are here conceived as rooms of light: an entire wall of glass opens the interior to beautiful sky-opened two levels secret patios with hanging gardens, flowers and tree species. The intimacy of each of these five emotional environments offers visitors a somewhat "comforting" experience of pain.

Each of the secret patios welcomes the work of a contemporary artist; the mortuary builds sites of affective sharing, spaces to live poetically thanks to the language of art. An outside "path" in height, through the mineral patio, leads to a mysterious outdoor belvedere, otherwise unreachable, facing the garden and beyond the profile of the Renaissance city walls. This special meditative space is designed for individual isolation and contemplation. Suspended just opposite to the transparent main front and facing the rising sun, the architectural body surrounds the courtyard, floating on air, embracing the visitors.

The old circular pre-existing pit is a great place to house a sacred tree, a universal symbol of life and rebirth in all cultural and religious beliefs. So the Citadel will celebrate death not as an interruption, but as a simple transformation of life. Thus, for this reason it will be called Domus Vitae, home for life.

二层 second floor

1. distribution
2. private entrance
3. private toilet
4. visitors' toilet
5. chapel
6. mourners' chamber
7. meditation space
8. prayer balcony
9. balcony
10. outdoor fire stairs

一层 first floor

1. corpse observation
2. infected corpse observation
3. technical plants
4. corpse conservation
5. infected corpse conservation
6. autopsy room
7. clinical findings deposit
8. analysis laboratory
9. clinical findings storage
10. corpse preparation
11. foyer and reception
12. visitors' toilet
13. multipurpose room
14. coroners' office
15. private service for male
16. private service for female
17. deposit storage
18. employees' offices
19. rest space/cafeteria
20. portico
21. garden and public space
22. outdoor patio

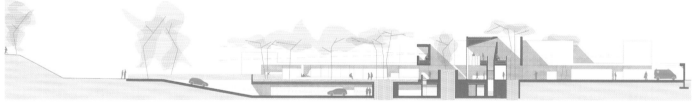

A-A' 剖面图 section A-A'

B-B' 剖面图 section B-B'

C-C' 剖面图 section C-C'

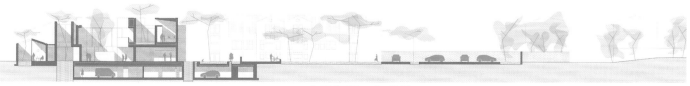

D-D' 剖面图 section D-D'

# St. Horto花园 _OFL Architecture

该竞赛获胜项目"St. Horto"的名字是一个双关语,由意大利语和英语两种语言组成。"Horto"和"St"这两个词都有一种神圣的私密花园的意思,同时意指它如同生长的植物一样,虽然有些弯曲和杂乱,却充满了生机。两者加在一起,表现出不规则的几何图形(弯曲或者歪歪斜斜),但是通过花园的建筑设计得到了高度控制。

"St.Horto"与项目区域完美地融为一体,同时通过压缩和扩张的空间设计手法,形成一个动态而富有吸引力的空间,重新定义了区域界限。这一设计理念的诞生源自于对纺纱方法的观察:机械的方法(毛纺厂)和手工做法(纺锤或卷线杆)。

这两种元素被再次诠释和结合到项目当中,以木质的柱子和白色的帆布绳索来表现。三角形花圃内种着各种各样的植物。三个不同的几何模块,重复组合,可以形成多种不同的三角形,营造出似乎无穷无尽的感觉。

为了与Lanificio的内在文化活动保持一致,建筑师们在花园内所采用的几何结构的造型来源于对多种艺术形式的模仿,包括绘画、图画、音乐和电影。

建筑的作用至关重要:所创造的空间不对称但又均衡,不均匀但又和谐,形成不同寻常的触觉、嗅觉和视觉等多种感官的冲击效果,为儿童们创造了完整的适合其身心发展的体验。

花园的入口位于屋顶的西侧,在露台入口的前面。

进入花园,你就进入了一个一连串充满动感的各种三角形形成的虚虚实实的空间当中,一个接着一个,沿着一条特别的富有教育意义的小路分布。

St.Horto的创新之处无疑是融合了为此项目专门打造的2.0技术。

受到其他项目中花园与传感器交互作用的启发,有了Jardimpu创造者阿尔贝托·塞拉的直接经验,建筑师们决定在花园中安装可以监控植物实时生长的传感器技术,使用硬件工具(带有传感器和网络摄像头的Arduino)和软件设施对花园里的植物进行监控。

他们在花园里三个特定的点使用了钢丝绳,并把它们设计成竖琴的样子。

St. Horto项目将功能性、美观性、生产性和教育性这四个元素紧密融为一体。生产性在设计布局中至关重要,在花圃中可用来栽培植物的区域有115m²(可灵活用于未来的扩建项目),因此足够用来进行真正的农业生产,而不只是单纯用于教育或训练目的。

## St. Horto

The name of the competition winning proposal "St. Horto" is a pun, made of Italian and English language. The two words "Horto" and "St" give at the same time the idea of a sacred, intimate garden and of something crooked, apparently disordered but full of life, just like growing plants. Together they represent the irregular geometry (crooked or oblique) but highly controlled architectural design of the garden. St. Horto fits perfectly within the project area and at the same time it redefines the boundaries through a game of compressions and expansions creating a dynamic and attractive space. The concept comes

cultivation-scheme

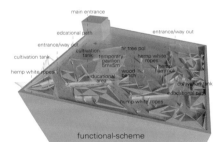
functional-scheme

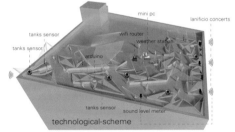
technological-scheme

to life from the observation of spinning methods: mechanical (woolen mill) and manual (spindle or mistaff).

These two elements have been reinterpreted and incorporated into the project and represented by wooden poles and ropes of white canvas. The triangular tanks containing the real garden are obtained by the repetition of three modules, which in combination create endless compositions. In line with the Lanificio's inner cultural activities, architects have chosen to adopt this geometry by the analogy with various forms of art: painting, drawing, music, and cinema.

The architecture plays a fundamental role: spaces which are asymmetric but proportionate, uneven but harmonic, create unusual tactile, olfactory and visual perspectives designed to facilitate a complete and suitable experience for children.

The access to the garden is on the west side of the roof, in front of the entrance to the terrace.

From here you can immerge yourself in a space characterized by a succession of energetic triangulations, full and empty, which follow one another along a path of particular educational functions.

The innovative feature of St. Horto is definitely its integration with the 2.0 technology through a customized project.

Inspired by other projects of interactive gardens with sensors and thanks to the direct experience of Alberto Serra, creator of Jardimpu, it was decided to install a technology allowing realtime monitoring of the growing plants in the garden, through the use of hardware tools (Arduino with sensors and webcam) and software.

In three particular points of the garden, they use steel cables which become veritable harp instruments.

The St. Horto project combines four inseparable factors among them: functionality, aesthetics, production and teaching. The production is essential in the designed layout. Inside the tanks the usable area for cultivation is 115 square meters (flexible for future expansions) and is therefore enough for a real farming production, not only purely educational/disciplinary.

项目名称：St. Horto
地点：Rome, Italy
建筑师：Francesco Lipari, Vanessa Todaro, Federico Giacomarra
项目团队：Mitchell Joachim, Alberto Serra, Nicola Corona, Felice Allievi
甲方：Lanificio 159 / Area: 200m²
设计时间：2012

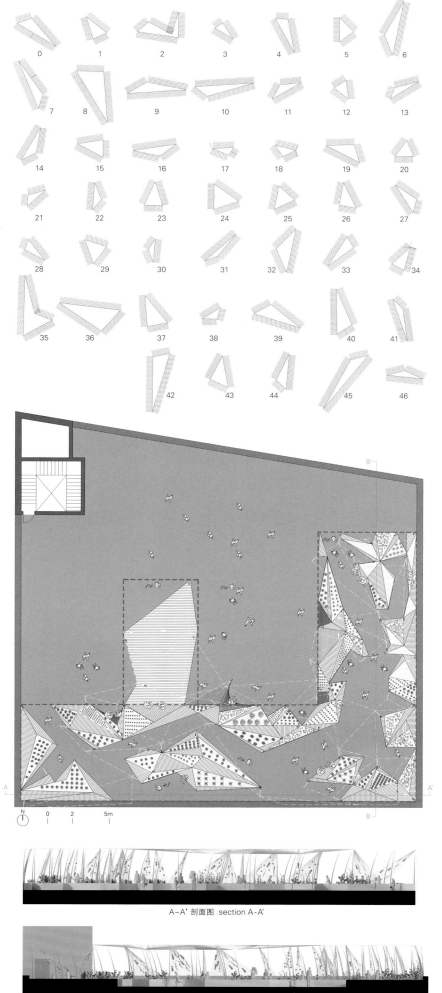

A-A' 剖面图 section A-A'

B-B' 剖面图 section B-B'

# 新加坡2012建筑节零浪费展馆 _Wow Architects

展馆的设计很好地呼应了项目所在地的二元性特征。一方面，Fort Canning山，曾被称作"禁山"，依然保持着安静、恬适、近乎神秘的特点。它的正对面是克拉克码头，充满活力，人流熙熙攘攘，活动丰富多彩。两者之间是Foothills，它曾经是一个热热闹闹的社交活动场所，有公共游泳池和国家剧院。

提案中的展馆力图体现这两个领域的二元性；建筑外表具有通透性。如波浪般起伏的网状结构让人感到惊奇的同时也能激发人们的好奇心。从某些角度看，这个网状膜结构看起来几乎像一面实体墙，由于该结构具有双覆层，当人沿着这面"奇迹墙"走动时，就会看到"莫尔条纹"效果。正面观看时，这个结构似乎完全透明，与周围的建筑和景观融于一体。

零浪费和建筑效益策略围绕可迅速开展和可再用这两个系统展开。第一个是主结构，由为一级方程式夜间赛车和国庆阅兵而开发的钢箱桁架系统组成。第二个是聚合物网格，为管理斜坡而开发，具有独特的属性，可以提高空间的可用性和交互性。网状膜结构及其景观系统可以被重新利用来维护加固Fort Canning山周围的斜坡，防止水土流失。建筑师的"零浪费"策略考虑到了时间、材料、成本和所有构件的展后处理。这个钢箱桁架系统，包括屋顶，最多需要约7天的时间来安装。网状膜结构最多需要约3天时间来安装。完成整面"奇迹墙"的架设所需的全部时间将是10～15天。

这个多孔的网状膜结构拆卸后，可以重新用于如下情形：

——用于Fort Canning山其他需要保护和加固斜坡的区域。

——捐赠给附近乡村，用于加固受到斜坡的水土流失所侵害的村庄/农田。

——钢箱桁架系统拆卸后不但可以用于将来的国庆游行，还可以重新用于其他商业活动。

网状膜结构通常被用作地下土壤保护技术，现在又被赋予一种新用途，即用作一个垂直平面，在上面进行设计、嵌入、与公众接触和互动。该结构成为一个展示平台，用来展示怎样用废弃塑料瓶栽培植物以及怎样应用零浪费策略。网状膜结构上面的孔洞成为与公众互动的"口袋"式亲密空间，成为人们传播思想和培育思想种子的地方。

——建筑节"明信片"在入口处发放给每位游客，鼓励游客在上面写下所思所想和美好记忆，然后放到网状膜结构上面的"口袋"里供所有人阅读、分享。

——展馆设计的另一个奇思妙想是不起眼的建筑节草席，用来营造公园气氛。草席卷起来插在"口袋"里，鼓励游客们坐在上面聊天、讲故事和分享彼此的经历。

——整个"奇迹墙"是一个城市小农场，塑料瓶里栽培着小观叶植物，植物根部用地质纤维包裹，并且将水凝胶植物栽培载体注入细胞之中，以保持水分。

## Archifest Zero Waste Pavilion

The design of the pavilion was a response to the duality of the site. On the one hand, Fort Canning, once known as the "Forbidden Hill" still retains its quiet, reposeful and almost mythical character. Directly opposite is Clarke Quay, vibrant and bustling with people and activities. In between is the Foothills that once was a hive for social activities with the public swimming pool and the National Theater.

The proposed pavilion seeks to embody the duality between the two realms, with its permeable skin. The undulating web inspires curiosity and amazement as well. At certain angles, the membrane looks almost like a solid wall, and when one moves along the Wonder Wall, a "moire" effect is created due to the double cladding around the structure. When viewed on the perpendicular, the membrane seems totally transparent and merges with the surrounding buildings and landscape.

The zero waste and buildability strategy was developed around two highly rapid deployable and re-useable systems. The first is the main structure, composed of box-truss systems developed for the Formula One night race and the National Day Parade. The second is a polymer mesh de-

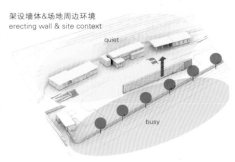
架设墙体&场地周边环境
erecting wall & site context

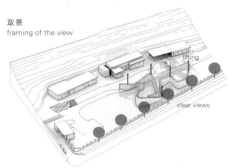
规划1&场地中各部分的关系
programming 1 & site relationship

取景
framing of the view

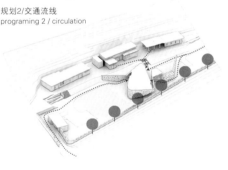
规划2/交通流线
programming 2 / circulation

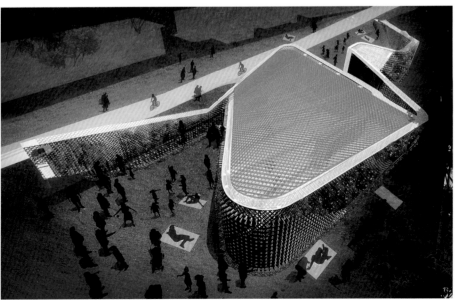

网状膜结构
membrane

donation to countries affected by soil erosion/slope protection at Fort Canning

structure

National Day Parade/commercial elements

瓶栽植物
plants

planting to Fort Canning

草席
mats

donation to third world conutries/used by people all over the world

明信片
postcards

ideas are documented in archifest website/postcards become recycled paper

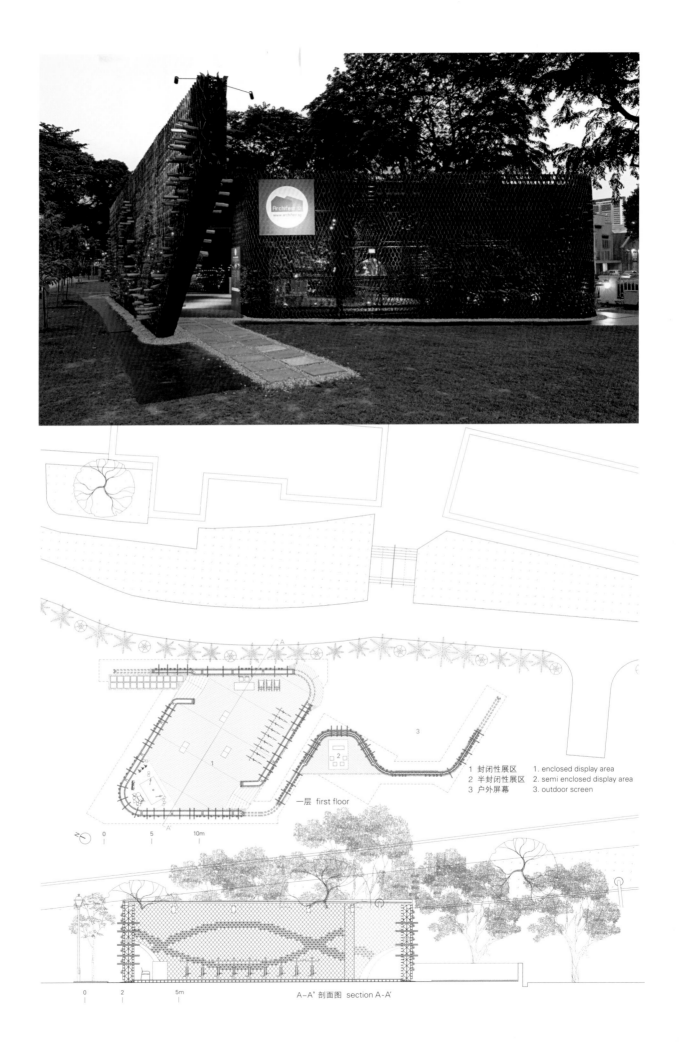

| | |
|---|---|
| 1 封闭性展区 | 1. enclosed display area |
| 2 半封闭性展区 | 2. semi enclosed display area |
| 3 户外屏幕 | 3. outdoor screen |

一层 first floor

A-A' 剖面图 section A-A'

veloped for slope control that has unique attributes that enhance the usability and interaction of the space, the membrane and its landscape system can be reutilized around Fort Canning for slope and erosion control. Our zero waste strategy considered time, materials, cost and the afterlife of the elements. The box-truss system, including the roof takes a maximum of approximately 7 days to delpoy. The membrane takes a maximum of approximately 3 days to install. Overall time frame to complete Wonder Wall erection would be 10-15 days.

The cellular membrane once taken down can be re-used for the following:

– Fort Canning hill's other areas that require slope protection and stabilization.

– Donate to a nearby country whose village / farmland has been affected by soil erosion from slopes.

– The steel box-truss once taken down will be re-used in other commercial events along with the future National Day Parades. Normally used as a subterranean soil control technology, the membrane is given a new use as a vertical surface onto which to project, insert, interact and engage with the public. Seminars on pop-up farming, and zero waste Strategies can be conducted using the Versiweb membrane as a display surface. The cellular nature of the mesh system also forms "pockets" of intimate space or crenellations in which seeds of thought are propagated and nurtured.

– Archifest "post" cards are distributed to visitors at the entrance and they are encouraged to post thoughts, ideas and memories in the "pockets" to be shared and read by all.

– One of the initial inspirations for the pavilion was the humble straw mat for a park-like atmosphere. Rolled up Archifest straw mats are inserted into the "pockets" to encourage visitors to sit and converse, tell stories and share experiences.

– The entire Wonder Wall is the urban pop-up farm with small foliage plants with geo-textile wrapped roots and a hydro-gel planting medium inserted into the cells.

项目名称：Archifest Zero Wastre Pavilion
地点：Singapore
建筑师：Wow Architects
用地面积：270m²
展馆总长：40.7m
网状膜结构墙体长：100m
高度：4.6m
材料：polycarbonate, wood-plastic composite(WPC) decking, veriweb 100 panels, recyclable PET, wedelias plants, straw mats, cable ties(nylon plastic )
竣工时间：2012
摄影师：©Aaron Pocock(courtesy of the architect) (except as noted)

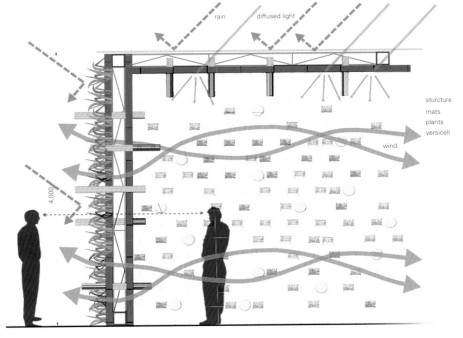

# 在城市中转换

## 建筑范围和城市影响力

　　在2013年的第一本书里，我们不禁自问，现如今城市里的建筑范围究竟有多大？一栋建筑物的运转影响范围有多大？某些建筑直接用正面展现其建筑特色，而另一些建筑会通过其与周边环境的关系来呈现。随着建筑与周边环境的物理关系范畴（视觉的、空间的和物质的）不断扩大，它们之间关系的性质变得越来越模糊和难以界定。对一些项目来说，很难说其影响的是建筑前面的大街、所在的社区，还是最终影响到了整座城市。

　　本书中我们所介绍的设计项目表现了对周围环境的不同处理方法，充分显示了建筑对城市环境形成的无形影响力。在"都市生活体验的场景"一章中，所有项目都穿插在密集的城市环境中，从而引发对城市更大程度的影响，涉及到城市的各个区域。根据Paula Melâneo的描述，某些建筑可以被视为"城市文化多样性的具体写照"。(p.28)

　　在"面朝街道，心系大海"一章中，我们看到建筑特色与城市街道有着千丝万缕的联系，通过观察建筑在城市全景中如何"显现"就可略见一斑。实际上，Diego Terna向我们解释了建筑是怎样"注重自身个性特点"，追求广泛的"认可"的。(p.98)

　　如果建筑对城市影响的范围扩大，受到无形影响的范围也会越大，更大的城市区域会被牵涉其中，建筑物与城市的关系也会变得更加紧密。为了更好地审视这一建筑处理方法，我们邀请了Simone Corda撰文反思城市设施的影响规模。Simone Corda在"重新思考地面状况"一文中描写了"建筑物与其环境"关系的模糊性，使其成为一个地形学问题(P.136)。

　　如果综合起来考虑，本书呈现给读者的设计项目描述了城市影响区域的变化，这种变化体现在通过直接和可见的功能和空间关系对单一个人的影响到对整个社区的影响。在社区中，建筑和城市之间所确立的影响和相互关系是无形的、间接的和模糊的。

# Variation and Transition

**Building Extent and Urban Influence**
With the first issue of 2013 we ask ourselves what is the extent of architectures for the city today. How much does the operative sphere of a building range in extents? Some building presents characteristics that relate it to its immediate front, other with its surroundings. As the physical relationship (visual, spatial and material) expands, so the nature of the relationship becomes increasingly sparse and undefined. For certain projects it is difficult to indicate whether their scale of influence is the front street, their neighborhood or eventually the city.
In this issue we present several projects which represent different approaches to their surroundings, showing how the invisible area of influence of a building may operate in the urban context. The projects of "*Scenario for Urban Daily Life Experience*" chapter are inserted in a dense urban context, which triggers a larger level of effect for the city: they relate to a modest 360° area. Paula Melâneo describes how certain buildings can be regarded as the "reflections of the city's cultures diversity". (p.28)
In "*Facing the Street, I am (not) a Duck*" we see buildings with architectural characteristics that relate them to the street, observed through the lens of their "emergence" in the urban panorama. Diego Terna – in fact – explains how they try to "bring attention to their own personalities" and strives for a broad "recognizability". (p.98)
If the scale of interference is increased, and the invisible sphere of influence becomes bigger, a larger urban area is called into question and the buildings relate to the city in a larger measure. To observe this approach, we asked Simone Corda to reflect upon the scale of the urban facilities in his article "*Re-thinking the Ground Level(s)*", which describes the blurring of "the architectural object and its context", becoming a topographical matter (p.136) .
If considered all at once, the presented projects depict the variation of the area of urban influence, from the single human scale with direct, immediate and visible functional and spatial relationship to the extent of the community, where the influence and the interrelationships established by architectures and city are intangible, indirect and blurred. Silvio Carta

城市设计 UrbanHow

# Scenario for Urban Daily Life Experience

# 都市生活体验的场景

飞人屋/Louis Paillard Architect&Urbanist
日本潮牌办公大楼/General Design
隐藏的住宅/Teatum+Teatum
阶梯住宅/Jaques Moussafir Architectes
狄维娜护理之家14/Estudio de Arquitectura Javier Terrados
巴洛克庭院公寓/OFIS Arhitekti
萨莫拉办公楼/Alberto Campo Baeza

都市生活体验的场景/Paula Melâneo

The Trapeze House/Louis Paillard Architect&Urbanist
Neighborhood Office Building/General Design
Hidden House/Teatum+Teatum
Step House/Jaques Moussafir Architectes
Divina Enfermera 14/Estudio de Arquitectura Javier Terrados
Baroque Court Apartments/OFIS Arhitekti
Zamora Offices/Alberto Campo Baeza

Scenario for Urban Daily Life Experience/Paula Melâneo

　　一般来说，社区是指与人、事物及地点临近或毗邻的地区或区域。本文将讨论社区居民及其周边环境。
　　在为建筑选址之前，建筑物的设计方案和用途是定义、定位、定性一座建筑物和社区关系的首要之事。
　　其次，必须评估建筑物的历史背景。建筑物在社区中已经发挥了作用？还是将会发挥新的作用？
　　然后，建筑地点明确了建筑物和周边环境的关系：建筑物是孤立存在还是见缝插针地坐落于人口密集的市区呢？是建在现有的建筑物之间呢？还是建在空地上？
　　完成这些工作后，可以分析建筑物的形状、材质或颜色是否和那一区域协调，该建筑物仅仅是个仿制品还是一个和所在区域的建筑根本不相融的怪物。
　　最后，建筑物的视觉效果、规模大小、内外体验讲述着建筑物和城市之间错综复杂的联系。

In general terms, Neighborhood refers to the notion of nearness or proximity of an area or space to people, things or places. In this analysis we talk about inhabitants and their immediate surrounding.
Even before choosing a site for a building, the first thing that defines, characterizes or constrains the building's relations with the neighborhood is its program and its use.
Secondly we must evaluate the historical context of the building: does it already play a role in the neighborhood or will it trace a new one?
Then, the insertion site clarifies its relation with the immediate surroundings: is the building isolated or inserted in a dense urban context? Is it built between existing constructions or in an empty plot?
After that, we can analyze if its shape, material or color will make it a reference in that space or if it will just be a mimesis or a discrete element of a set.
Finally, the visual performance of the building, its size and scale, and its experience inside out or vice versa, can also explain the complexity of the connections with the city.

社区是城市文化的产物。社区风格迥异,恰是城市文化多样性的反映。每个城市都有自己独特的文化特征,取决于诸如地理位置、理念、模式、政治、决策、价值观、时间、距离等不同条件以及复杂因素。这些制约形成了城市文化的特征,而这些特征不同程度地影响着人们的生活方式。从城市的宏观角度看,我们说的是城市公民的生活;从社区这一较小角度看,我们考虑的是每个个体的生活。

社区的建造通常是为了满足在某一时间段内人们对建筑物的特殊功能或文化需求,这一特性可以决定或形成某类风格,这就是通常所说的社区某种程度的单一性。社区结构或分散或紧凑,条条街道形成了一个多多少少有些复杂的结构,从而成为城市复杂关系网中的初级单位。

主要的问题是人们在这些社区内如何交往,人际交往需要怎样的条件以及他们将怎样运用这些条件。社区是城市日常生活体验的主要背景,城市文化这一"剧本"在社区中上演。通过日常经历和空间活动,个体建构出自身情感、情绪、记忆、活动地点的个人数据库。

简·雅各布斯在其1961出版的《美国大城市的死与生》一书中特别阐明,安全感和幸福感是居民在城市和社区生活中最重要的品质。雅各布斯说道,安全感和幸福感可以在繁忙拥挤的空间中获得,而与之相关的社会关系能确定一个生活空间的品质。这就质疑了现代城市的规划、过度设计的城市或没有历史底蕴的城市,这样的城市是与居住个体毫无关联的贫瘠之地。

如今,我们所处的社会似乎渐渐摈弃面对面/私人交往,邻居对我们而言不过是陌生人罢了。我们不禁要问,将来社区和城市中的生活会是什么样,人们往往会说这么做是为了保护个人隐私,但为什么与此同时,人们却将自己的生活完全暴露在网络上?

根据其建筑风格在临近社区产生的社会和空间影响,下文将逐一分析这些精选工程的主要特点。

如上所述,分析的第一步是设计方案和用途。除了公用建筑外,精选出的建筑全部是私用或半私用的工程。有些工程设计只用来办公,有些用来居住,还有一个工程将居住与办公合二为一。此外,建筑

The Neighborhood is a product of a city culture. They are diverse as they are reflections of those cultures' diversity. City culture is not homogeneous among cities: it depends on such different conditions and complex factors as geography, ideas, models, politics, decisions, values, time, distances, etc.. These constraints induce to the construction of identities of city cultures that can influence ways of life in different scales. In the global scale of the city we speak about citizens' life and in the smaller scale, of the neighborhood, we consider individuals' life.

A neighborhood is normally built to meet a specific function or culture within a time lapse and this can determine or characterize a style. This leads to a certain homogeneity that often defines it. Disperse or compact, with streets defining a more or less complex fabric, the neighborhood acts as the first level of network relations complexity in the city.

The main questions are how people interact in those neighborhood spaces, which conditions they have for that, and how they will appropriate them. The neighborhood is the primary scenario of the daily urban living experience, where city's culture acts take place. These everyday experiences and spatial actions, allow individuals to construct their repertory of affections, emotions and memories, their own personal references to physical places.

Jane Jacobs (*The Death and Life of Great American Cities*, 1961) highlighted safety and the feeling of well-being as some of the most important qualities in city spaces and neighborhoods. She mentioned that these conditions could be found in busy and crowded spaces, the social relations being relevant to define the quality of the space. This questions modern town planning, over-designed cities or cities without history, that could be considered as sterile spaces without references to individuals.

Nowadays, when society seems to be abandoning face-to-face/personal relations, when our neighbors are nothing more than strangers to us, we question what the future of urban life experience in neighborhoods and cities will be. Why this trend is to seek a pretended physical privacy when people expose their entire life on the Internet?

The following selected projects are analyzed according to their main characteristics that can define the social and spatial influence of a building's architecture in its close neighborhood.

As mentioned, program and use are the first step to the analysis. Excluding buildings for public use, this selection presents exclusively projects that are for private or semi-private use. Some have a program entirely related to work, others to housing and one has a mixed program. Then, the historical context, the insertion site,

隐藏的住宅,位于典型的砖石建筑之间
Hidden House, positioned in between the typical brick buildings

照片提供:©Teatum + Teatum (Lyndon Douglas)

的历史背景、建筑地点、建筑的自身特点和各自的视觉效果将会为我们展示建筑本身是如何在某一特定社区发挥作用的。

**复古风格**
**狄维娜护理之家14,Javier Terrados建筑工作室设计**

这座新建筑计划修建小型公寓以出租给老人和青年人。修建伊始,狄维娜护理之家14已经改善了不同代际人们之间的社会关系。

狄维娜护理之家14修建在塞维利亚市中心,建筑场地为市中心仅有的几块空地之一。此建筑采用了以庭院为中心的住宅结构,类似这一传统西班牙历史名城中"院落式"的居住方式。在这里,居民们聚会或进行社交活动,远离城市无止的喧嚣,也免受塞维利亚夏季骄阳的曝晒。

如同建筑师所说,尽管庭院具有半私密的性质,它在某种程度上却成为街道的延伸部分。在数字化交流的社会,这一建筑勾起了人们对以往面对面交流的怀旧情怀。在这里,居民可以在共享的内部庭院聚会,在开放式走廊里走来走去,碰见自己的邻居,每天都有新鲜事发生。

在博客中,Terrados思考了赫曼·赫茨伯格的话并指出,在建筑学中需要着重考虑的往往是建造在"建筑之间"的物体,但是"建筑之间"恰好是让人难以区分什么是内、什么是外的区域。

尽管这座建筑位于一个封闭的场地,从街道难以入内,且在城市里很不起眼,却很好地展示了社会交往是如何在私人建筑中蓬勃开展的。

**黑色仓库**
**日本潮牌办公大楼,General Design设计**

这栋混凝土建筑位于涩谷(东京23个区之一)。在涩谷,购物、时尚秀、夜店在年轻人中很受欢迎。该大楼即身处这样的环境中,其内有一家服装公司,外带办公室和一间展览室。区分这种双重功能很容易,大楼外部有两个通道进入公司,分别位于大楼的两侧;一个直通工作间,另一个直通展览室和后勤区。

大楼外部的深色调和封闭的亭状结构使整座建筑看上去像工业建筑,而这正是该建筑的特征。大楼的外形是其身份标志,使它在社区内很容易辨认。尽管它是封闭建筑,却因为分布于不同楼层的两个

the physical characteristics and respective visual performance will show us how the building operates in a specific neighborhood.

**Like in the Old Times**
**Divina Enfermera 14 by Estudio de Arquitectura Javier Terrados**

The program for this new building claimed for small area dwellings to rent to elder people and young adults. As a start, it already promotes the social relation between different generations.

It was built in one of the last empty interior plots in Seville's center. This location inspired a central courtyard solution as the traditional "corral de vecinos", a residential system of the historic city. Here neighbors can meet and socialize, protected from the constant noise of the city and from Seville's hot sun, during the summer.

As explained by the architect, the courtyard somehow functions as an extension of the street, even if it is a semi-private typology. In a digital communication society, this building represents certain nostalgia for the face to face social relations. Here people can gather in this common interior courtyard and circulate across the open corridors, meeting their neighbors and living a new story everyday. In his blog, Terrados remembers Herman Hertzberger's words referring that the important thing to consider in architecture is what happens in the "in-between", exactly the place which is difficult to distinguish between inside and outside.

Even if this building is located in a closed plot, with difficult access from the street, and with a discreet presence in the city, it's a good example of how private architecture welcomes the social relations into a building.

**The Black Warehouse**
**Neighborhood Office Building by General Design**

This concrete building is located in Shibuya, one of Tokyo's 23 wards, where shopping, fashion and night entertainment are popular among young people. In this context, the building hosts a clothing company, with offices and a showroom. Dividing this double function is made easier by its double access from the exterior, by opposite sides. One of the side entries leads to the workspace while the other gives direct access to showroom and logistics area.

The exterior dark color and closed pavilion structure, just like industrial architecture, give it character. It has an image that can work as a reference, allowing it to be easily identified in the neighborhood. Despite being a closed building, it has a dynamic relation with the street, with its double access at different levels.
In the interior, a central skylight lightens the unvarnished environment. The openness of the space enables the interconnectivity between the floors and the workspaces that allow a closer

飞人屋,嵌在两座相邻的建筑之间
The Trapeze House, inserted in between the two contiguous buildings

入口而与街道形成动态联系。

大楼内部的中央天窗照亮了不加装饰的四壁。空间的开放性使得各楼层和各工作间相互贯通,也使员工之间关系更加紧密,建筑师称,这种设计提高了员工的创造力和生产力。

**内外混淆?**
**阶梯住宅,Jacques Moussafir建筑师事务所设计**

在巴黎,许多社区常常"隐身"于城市街区内。这一独栋私人住宅坐落于巴黎市正中心,位于一个典型巴黎式"庭院"内,但从街上却完全看不到。这栋房屋犹如一个私密的过滤器,只有隔壁的邻居才能看其面目;而从外面看来,这栋房屋与城市是没有任何联系的。

这座亭阁式住宅南侧的玻璃立面让人想起古代花园中带金属结构的古老玻璃亭。该住宅与邻近的建筑物并置,同时和对面的建筑物距离适中。尽管南立面完全透明,能看到内部的天井,但金属百叶窗可以调节人们的视线。玻璃砖遮挡了望向住宅后侧的视线。内部结构沿屋内中央的楼梯螺旋上升,每一层都有公共开放区,使各楼层和各房间之间保持空间连续性,并使居住者之间关系更亲密。

**直面天空**
**隐藏的住宅,Teatum+Teatum建筑师事务所设计**

第一眼看见这个房子,人们不禁疑惑,这是否真的是一栋房子,因为它没有窗户,看上去更像个仓库。外立面是一堵厚重的墙体,有两层楼高,俯视屋外的辅道,底层入口安装有一扇玻璃双重门,外辅有一扇钢质门。隐藏的住宅夹在伦敦市中心后侧的建筑群中。在原有的典型砖石建筑中间,封闭外墙的黑色调和材质使房屋格外显眼,然而令人称奇的是其明亮的室内:整座住宅没有一扇窗户,而是开向天空,通过天窗采光井来照亮室内。隐藏的住宅在社区内引人注目,但同时可以保证居住者的日常生活有良好的封闭性、私密性和亲密性。

**光影交叠**
**巴洛克庭院公寓,OFIS建筑师事务所设计**

在坚固而古老的卢布尔雅那城(斯洛文尼亚首都)中心,OFIS建筑师事务所采用的新式手法是将三个巴洛克街区老宅重新排布,并最

relation among the team, for the benefit of creativity and production, as explained by the architects.

**Inside or Outside?**
**Step House by Jacques Moussafir Architectes**

Often in Paris, many city blocks "hide" neighborhoods in their interior. This singular private house is completely invisible from the street. It stands inside a typical Parisian "cour", in the city very best center. This system acts as a privacy filter where just the direct neighbors have a visual experience of the house. The visual connection with the city is nonexistent!

The glazed south facade of this pavilion-like house reminds us of the old glass pavilions, with a metal structure, of the ancient gardens. The house is juxtaposed to the contiguous buildings and has a nice distance from the facing buildings. Although the south facade is completely transparent to the interior patio, this relation can be controlled by metal blinds. The visual relation with the rear side is blocked by glass bricks. The interior evolves in a spiral around a central core with stairs, with open areas at different levels, which allows a spatial continuity between the floors and the spaces, and a closer relation between the residents.

**A House Opened to the Sky**
**Hidden House by Teatum+Teatum**

At first glance of this project, it made us wonder if it really was a house, because it has no windows; it could also be a warehouse. The exterior facade, over a secondary street, is a massive wall, two stories high, with an entrance glass door in the bottom, hidden by a steel second door. The house exists in an interstitial space of a London's urban center rear. Between existing typical brick buildings, its closed facade stands out because of its black colour and materiality. And the surprise is the bright interior. Open to the sky, this house with no windows is lightened by skylight wells.
The house is very identifiable in this back neighborhood, but guarantees to the dwellers a very closed, private and intimate daily use.

**Image Overlays**
**Baroque Court Apartments by OFIS Arhitekti**

In the consolidated and old city center of Ljubljana, this renovation by OFIS Arhitekti is a good example of reorganizing the space of three different baroque blockhouses, into a single unit with 12 apartments, as explained in their text. This operation unifies the space, enabling one general circulation for optimizing the space use. An interior courtyard serves as a common space for the three buildings, where horizontal circulation is used to connect the apartments. Depending on the light, this glazing courtyard is transparent, allowing daylight to penetrate the court, the circu-

萨莫拉办公楼,外围有高大厚重的石墙,但完全开向天空
Zamora Offices, enclosed by the massive walls but completely open to the sky

终将它们连为一个拥有12个公寓的统一整体。这一做法将空间整合起来,创造出一个总的交通环线,从而尽可能地利用空间。内部庭院是这三座老宅的公共空间,庭院中的一条水平走廊将公寓串连起来。在阳光照射下,该玻璃庭院完全透明,阳光透过庭院,穿过走廊以及各间公寓,有时光线反射四散,光影交叠,美不胜收。

### 合二为一
### 飞人屋,Louis Paillard Architect&Urbanist设计

这所房屋的独特之处首先体现于自身的设计:它为法国巴黎东郊蒙特勒伊的一个家庭提供住宅和工作室,练习马戏团飞人技巧。

住宅区位于一层和二层,几何外形很普通。顶楼露天,高9m,用于练习空中飞人技巧。

建造在两座风格迥异的建筑物之间,这座新房子自身风格又再度变换。建造方法、建筑形状、建筑材料将房屋职能划分开来。较低的混凝土楼层外覆有木板条,顶层是金属结构,覆有金色半透明的聚碳酸酯面板。顶层工作室有一个独立的出口直通街道,以保证在空中飞人训练课期间,生活和工作互不干扰,充分享有家庭隐私。飞人屋后侧有私人的外部活动空间,而从客厅可以直接通到内部小花园。

飞人屋非常吸引人的眼球,同时也是这一寻常城郊社区中其他建筑的参照物。

### 围墙里的办公楼
### 萨莫拉办公楼,Alberto Campo Baeza设计

这座新建筑与萨莫拉罗马式教堂和广场位于同一场地上。它的石墙围绕场地一周,石墙美观、厚重且高大;该建筑只有极少的开口可以看到街道和社区,它没有屋顶,直面天空。这让人不禁揣摩,它是军事堡垒还是一座私人城堡。

Campo Baeza称自己的设计为"Hortus Conclusus",即"封闭花园",它带有中世纪基督教色彩,注重精神情调。如今,它已不仅是一个花园,而是一座半私人化的石质院落,树木点缀其间。围墙内矗立着一座玻璃办公楼,它在外面看来是完全透明的,从而使这座植有树木的半私人化石质院落和内部的办公区可以进行直接互动。

lation and the apartments, or it is reflective, and the images are multiple, in magic overlays.

The facades to the street retrieved their original state. Not far from the castle hill, the roof floor has a very strong visual connection with the city roofs and a close relation with the castle.

### Two in One
### The Trapeze House by Louis Paillard Architect & Urbanist

The uniqueness of this building is in the first place demonstrated by its own program: a house and studio for a circus trapezist in Montreuil, a city in the Paris' periphery.

The dwelling space is located in the ground and first floor, and has a common geometry. The top of the building is a 9 meters high open space for the trapeze practice.

Built between two contiguous buildings that are already too diverse, also this new house imposes the difference. The construction method, shape and materials, separate the functions of this house. The concrete lower floors are covered in wood stripes and the top is a metallic structure covered with golden translucent polycarbonate panels. This top studio has an independent entrance from the street, as to guarantee the privacy of the house, separating work and family life during the trapeze lessons. The house has its own private exterior space in the rear side, an internal garden accessed directly by the living room.

The Trapeze House is certainly a visual mark and a quality reference on this ordinary suburban neighborhood.

### The Walled Office
### Zamora Offices by Alberto Campo Baeza

In the same square of the Zamora's Romanesque Cathedral, this new construction offers the city an image of a beautiful, thick, high and massive wall that outlines the entire site. With a few openings that frame street and neighborhood alike and without roofing, this new structure is completely open to the sky. We could ask ourselves if it is a military construction or just a barrier for a private property.

Campo Baeza calls his project "Hortus Conclusus", which means "enclosed garden" in Latin and it was a medieval theme with a spiritual significance, related to Christian religion. But today, more than a garden this is a semi-private stone patio with some trees. Inside the exterior walls, stands a glazed office building which by being totally transparent to the exterior creates a direct interaction between the semi-private patio with trees and the work spaces inside. Paula Melâneo

# 飞人屋

Louis Paillard Architect & Urbanist

该住宅项目有起居和工作两种功能。

该住宅位于巴黎东郊蒙特勒伊市中心附近，建在一块112m²的梯形土地上；房高15m，是这座城市所允许的建筑物最高限度。它有三层（其中一层是地下室），顶层为9m高的大型活动室，用来练习空中飞人技巧。

住宅沿街而立，一侧紧挨由黑砖砌成的住宅，其橙色窗框也由建筑师设计，另一侧临近一所巴黎市郊典型的磨石住宅。建筑面朝西侧，所有的开口都按照室内空间大小设置，以便从街区和社区中央的小花园引入充足的自然光线。

两个不同却经济实惠的建造系统凸显了住宅的两种功能：起居部分是深色的，有一个混凝土基座，垂直方向上带有锐利的立面；而顶部的训练空间则覆盖着金色聚碳酸酯面板，框架外包有金属，金色与暖铜色木质临街面色彩搭配和谐。

室内没有隔间也没有门，起居区域是重叠的三层，天花板向下倾斜，整个住宅结构是开放式的，不使用横梁，而是适当地在屋角切出几个空间，使其在空间上相呼应。

地下室是儿童游乐区，那里是儿童自由自在玩耍的乐园。一层有车库，通向厨房，还有一个宽敞的休息处，通向花园；一层由宽阔的水平推拉窗照亮。二层有侧边花园和父母的卧室，卧室沿着水体延伸，直到休息处空闲区的玻璃墙。在临街面，整个空间被划在一侧，有一间垂直于住宅正立面的图书馆和儿童的小卧室，横杆上挂着吸声窗帘，隔开外界的喧杂。

四个楼层由右侧两部水平交叉的混凝土楼梯进行服务，一个楼梯直接从街道通到活动室，这是为了在举办活动时保护家人的隐私。

活动室单一的体量比例精确，几何样式上参考了摆动的软梯呈现的弧线。大梁沿对角线搭建。这样住宅结构和屋顶覆层就向上拱起来，保证空中杂技师在表演时不会碰到屋顶。

出于"美学经济"的考虑，室内居住区给人随意、粗糙之感，地面由混凝土铺就，内埋有电热管，天花板和楼梯也显得很粗糙；与此相对，几堵墙壁则内衬有白色石膏板。唯一的例外是在表演室，墙上的壁画是由彩色玻璃镶嵌的器皿组成的，最终形成了户主的马赛克图像。表演工作室与它的设计理念相类似：钢质结构外涂防锈灰漆；齿轮墙和楼梯出口处的三角形漏斗结构只涂抹了灰色的水泥；裸露在外的电缆架；搪瓷夹心钢板屋顶覆层；混凝土铺就的供暖地面。只有宽大的镀铝窗镶嵌在聚碳酸酯面板上，与周围的景观协调一致。

这座独栋住宅向内深入挖掘的做法展现出一种崭新的现代建筑类型：本土风格和都市风格相交映，自然材质（起居区的木材）和人工材质（表演工作区的塑料板）相对比，同时呈现了粗糙和精细两种建筑美学。

人们是不是可以说这样一座建筑不是时代的建筑，而是另类的"城市谷仓"呢？

## The Trapeze House

The program of this house consists of two functions: to live and work in the same place.

Located near the downtown of Montreuil (in the east of Paris) in full rebuilding, the house was built on a small trapezoidal ground of 112 m². Built on the totality of the ground it rises 15m on top, the maximum height authorized by the payment of the city. It gathers vertically on 3 levels (of which one in basement) spaces of dwelling and is elevated on the last level by a high gymnasium of 9m dedicated to the practice circassienne of the trapezoid.

Aligned on the street, the house is flanked of an operation of residences in dark brick and the framings of orange windows are also carried out by the architect while being next to, the other side, a small typical house of Parisian millstone suburbs. Because its orientation is western, the whole of the openings, dimensioned

according to interior spaces, take the natural light on street and a small garden in the heart of a small island.

The two functions of the house are marked by two different but economic constructive systems: the dwelling part makes up of a concrete base barded vertically of blades, and the Douglas one, the space dedicated to the trapezoid is carried out in covered metal framework of sandwich panels out of polycarbonate gold to harmonize itself with the frontage out of wood treated with copper.

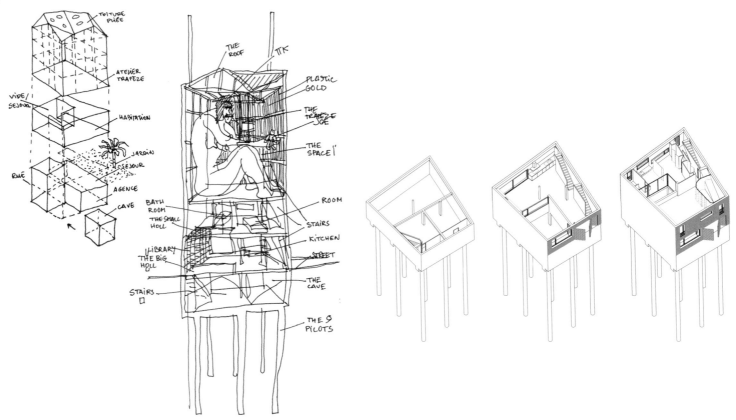

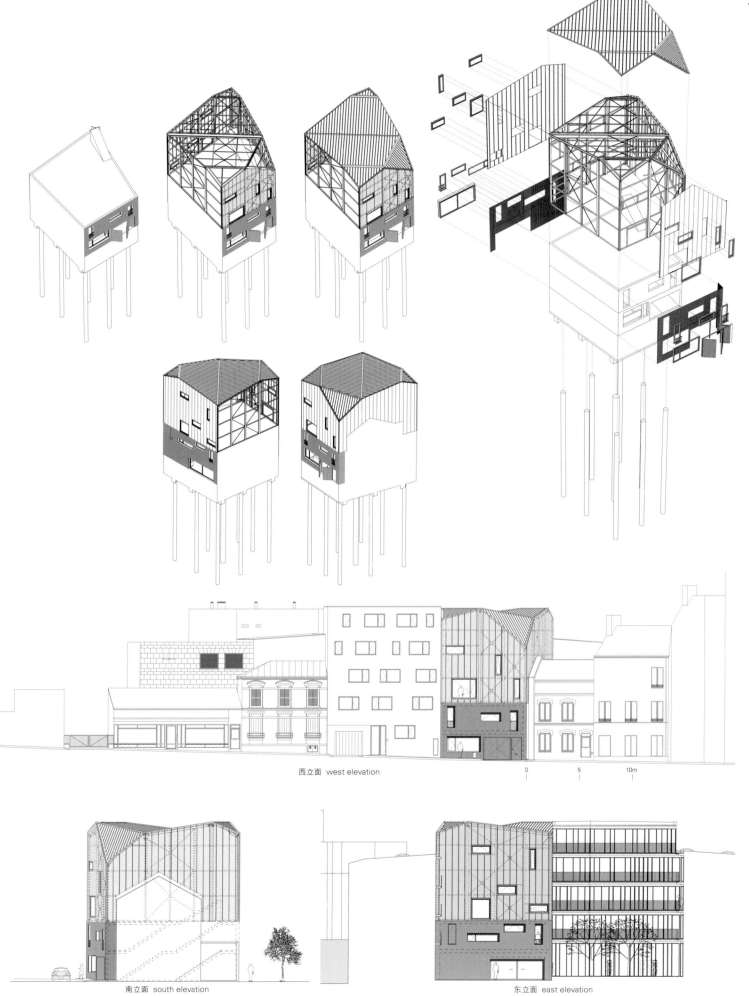

西立面 west elevation

南立面 south elevation

东立面 east elevation

Spaces without partitions nor doors, of the dwelling part superimpose on three plates with the slipping – by ceilings, left gross dismantling without repercussion of beam, and simply opportunely cut out vacuums which are answered the ones and the others spatially.

Under ground of the loft type is reserved to the children in the form of a vast freely appropriable space. The ground floor contains the garage open on the kitchen and a great stay turned towards the garden. The whole is lit by a broad sliding horizontal picture window. The first stage gathers side garden and the room of the parents prolonged by a vast water part to the glazed walls giving on the vacuum of the stay. Side street, the whole of space is divided on a side by an office library perpendicular to the frontage and of a small room of children to the acoustic curtains sliding to the rail of the ground to the ceiling.

The four plates are served laterally by two right concrete staircases which are superimposed; one carries out directly in two flights to the gymnasium since the street is to spare the intimacy of the family when courses of trapezoid are given there.

The singular volumetry of the gymnasium was precisely proportioned and drew geometrically according to the maximum race in curve of the swinging trapezoid. Thus the structure and the cover of the roof were folded so that the trapezist's moving never touched the roof.

For preoccupations "with an aesthetic economy" interior spaces of the dwelling part presents voluntarily rough completions: covering concrete with electric heating floors, ceilings and staircases rough of dismantling. By contrast the vertical walls are doubled in white plasterboards. The only exception, the fresco "warholienne" of the water room made out of earthenware of colored glass represents a pixelized photograph of the owners. The workshop trapezoid is carried out in the same spirit: structure of the steel frame painted in anti-rust gray; pinion walls and triangular hopper of the exit of staircase left in coating gray cement; apparent electric cable shelf; cover of the roof in industrial sandwich panels out of enamelled steel; cover of concrete with heating floor. Only large gilded aluminum windows inserted into naked polycarbonate tally the surrounding landscape.

This monolithic house as dug interior tries to produce a new contemporary architectural typology, between vernacular and urban abrasive by contrast of materials natural (the wood of the habitat part) and artificial (the plastic of the workshop trapezoid) which at the same time produces a rough and sophisticated esthetics.

Can one say that this house is not an in-time house but rather a kind of "urban barn"? Louis Paillard Architect & Urbanist

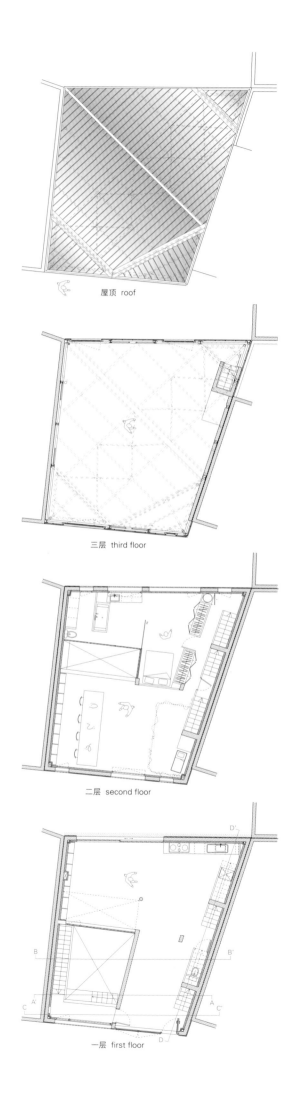

屋顶 roof

三层 third floor

二层 second floor

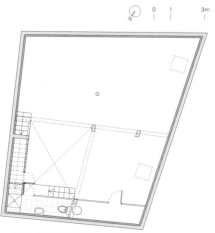

地下一层 first floor below ground

一层 first floor

项目名称：The Trapeze House
地点：6/8 rue du 18 Août, 93100 Montreuil, France
建筑师：Louis Paillard Architect & Urbanist
项目主管：Nicolas Lebatard
甲方：Paillard Familly
用途：house, trapeze workshop
用地面积：112m²
总建筑面积：350m²
建筑规模：one story below ground, three stories above ground
造价：EUR 650,000
竣工时间：2009.2
摄影师：©Luc Boegly (courtesy of the architect)

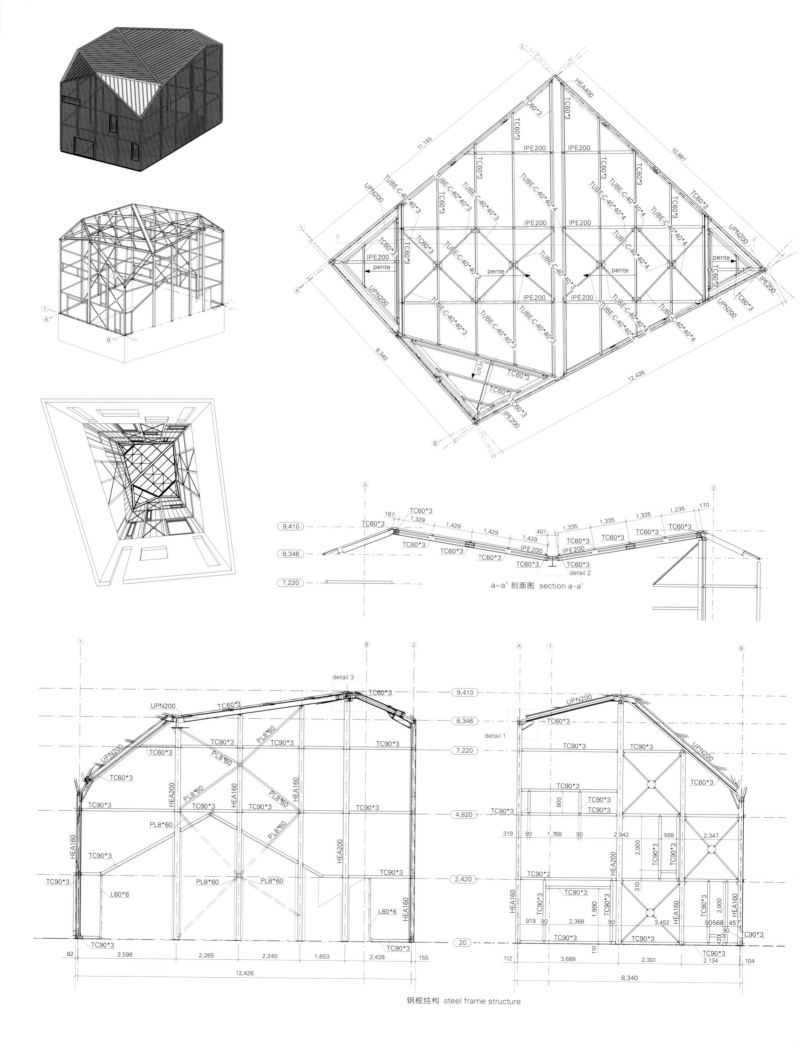

钢框结构 steel frame structure

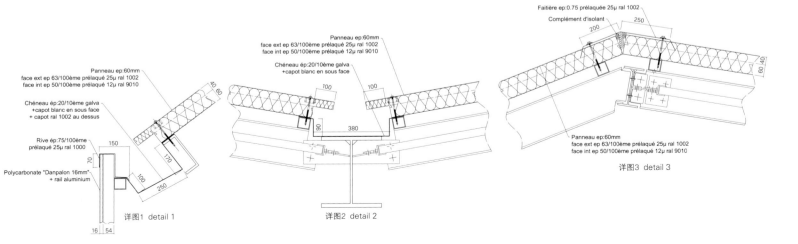

详图1 detail 1  
详图2 detail 2  
详图3 detail 3

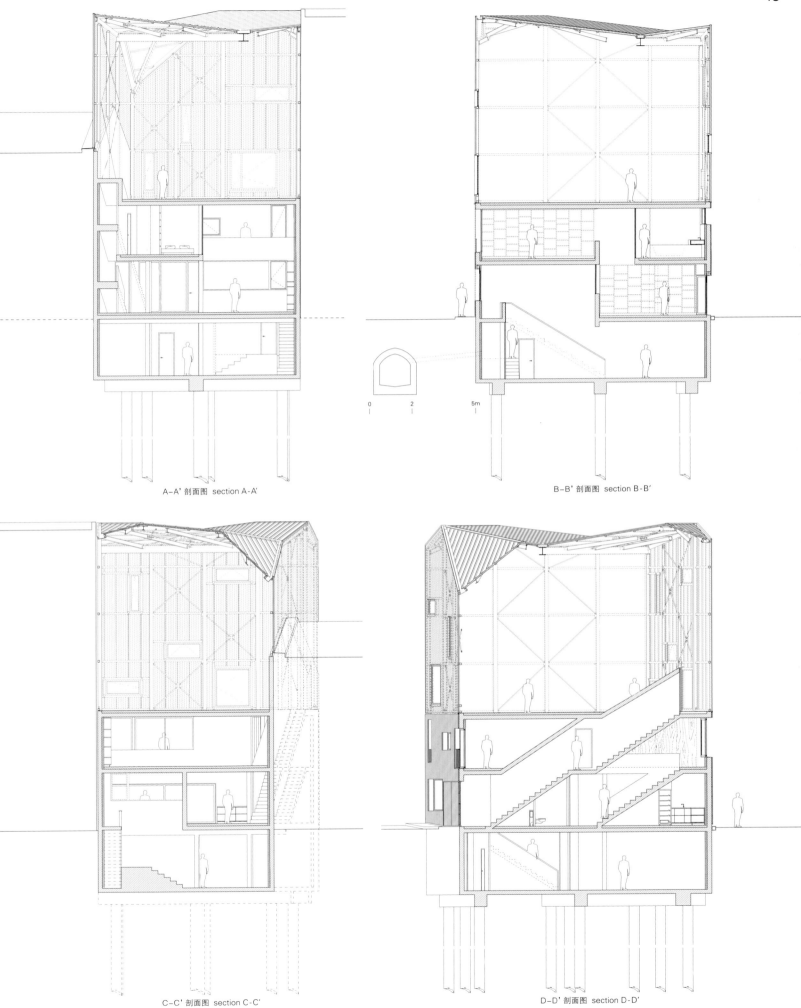

# 日本潮牌办公大楼

General Design

日本潮牌办公大楼位于东京涩谷区，是一家服装品牌公司总部大楼。公司生产大众服装，受到与摩托车、军事、户外运动等相关的多种亚文化的影响。

大楼的一层和二层是工作区，车库和展览室设在地下各层。

由于公司的员工相对较少，所以公司优先考虑的不是将办公空间最大化。与传统的办公楼设计——将一整层楼都用作办公区，在后部设置垂直交通流线——不同，这座办公大楼的设计使各楼层的工作区都有更多的互动空间，它旨在改善工作环境，提高员工的创造力和生产力。

大楼中轴上安装有长长的天窗，沿着大楼中轴，是一排细长的采光井。不同楼层的工作间全都位于采光井的两侧，由沿着中轴的倾斜楼梯相连。采光井在确保各个工作间之间保持适宜距离的同时，也为员工创造了一种集体感。

建筑师想要将办公空间建造得像仓库一样简洁而朴实无华。他们选用的材料是粗糙的露石混凝土、纤维钢筋水泥板、黑漆钢、落叶松胶合板，均选用工业建筑使用的普通原材料，尽可能减少建筑的"新"感觉，消除新建筑构件给人的突兀感。

本案建筑师希望这座办公大楼最终会自然变旧，就像一件珍藏许久的珍贵老式服装，随着时间流逝变得更有魅力。

他们的目标就是设计出未来的老式建筑。

## Neighborhood Office Building

This is a head office building of a Tokyo-based apparel brand "Neighborhood" which offers basic clothing influenced by various subcultures associated with motorcycles, military, outdoor etc., located in Shibuya district in Tokyo.

The building consists of workspaces on the first and the second floors, garage and presentation room on the basement floors.

Since the company has a relatively small number of employees, it was not top priority to maximize office space. Instead of typical office building plan, using entire floor area for office space with vertical circulation located in the back, we decided to allow more space for interconnectivity between workspaces on all floors, to enhance creative and productive work environment.

Long skylight is provided along the central axis, creating a sort of long and thin light well along the central axis. All workspaces are located at different levels on both sides of the light well, and they are connected with gently inclined stairs running along the central axis. The light well helps them to keep comfortable distance between all workspaces, while creating a sense of togetherness at the same time.

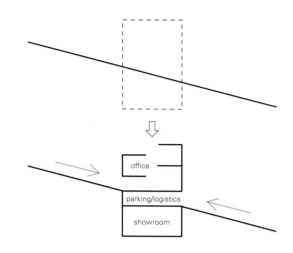

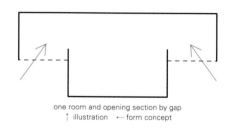

one room and opening section by gap
↑ illustration ← form concept

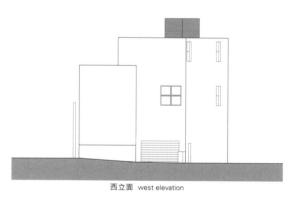

西立面 west elevation

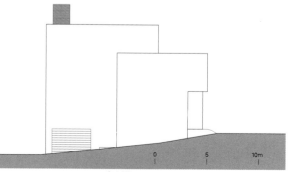

东立面 east elevation

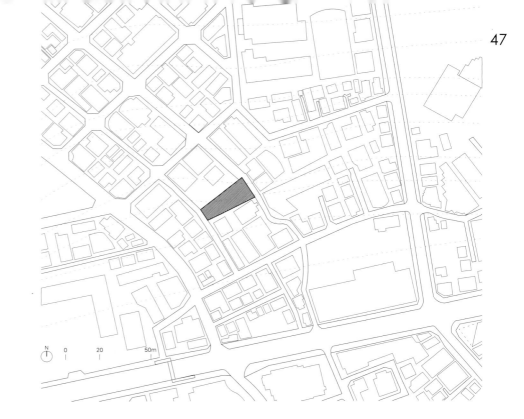

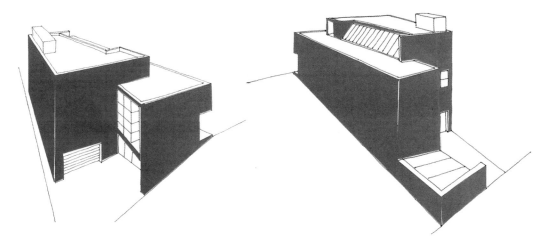

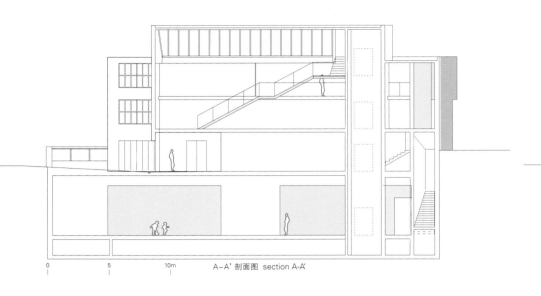

A-A' 剖面图 section A-A'

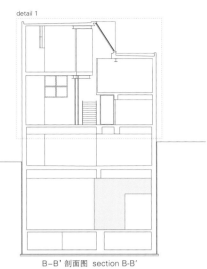

B-B' 剖面图 section B-B'

1 办公空间 2 屋顶花园
1. workspace 2. roof garden
三层 third floor

1 办公空间 2 储藏室 3 会议室
1. workspace 2. storage 3. meeting room
二层 second floor

1 办公空间 2 入口大厅
1. workspace 2. entrance hall
一层 first floor

1 储藏室 2 车库
1. storage 2. garage
地下一层 first floor below ground

1 展览室 2 办公空间
1. exhibition space 2. workspace
地下二层 second floor below ground

We intended to create a rather bare and unembellished space like a warehouse. Finish materials we chose to use are rough exposed concrete, fiber reinforced cement board, black painted steel and larch plywood. We selected ordinary and rough materials generally used in industrial environments, in order to minimize the sense of "newness" and eliminate emphasized strong presence of each element.

We hope that this building will eventually be nicely worn out and be even more attractive as time goes by, like a piece of good vintage clothing that you've cherished for a long time.

Our ambition is to propose future vintage architecture.

General Design

项目名称：Neighborhood Office Building
地点：Shibuya, Tokyo
建筑师：Shin Ohori
结构：reinforced concrete
用地面积：402.17m²
建筑面积：236.98m²
总建筑面积：992.31m²
摄影师：©Daici Ano(courtesy of the architect)

项目名称：Neighborhood Office Building
地点：Shibuya, Tokyo
建筑师：Shin Ohori

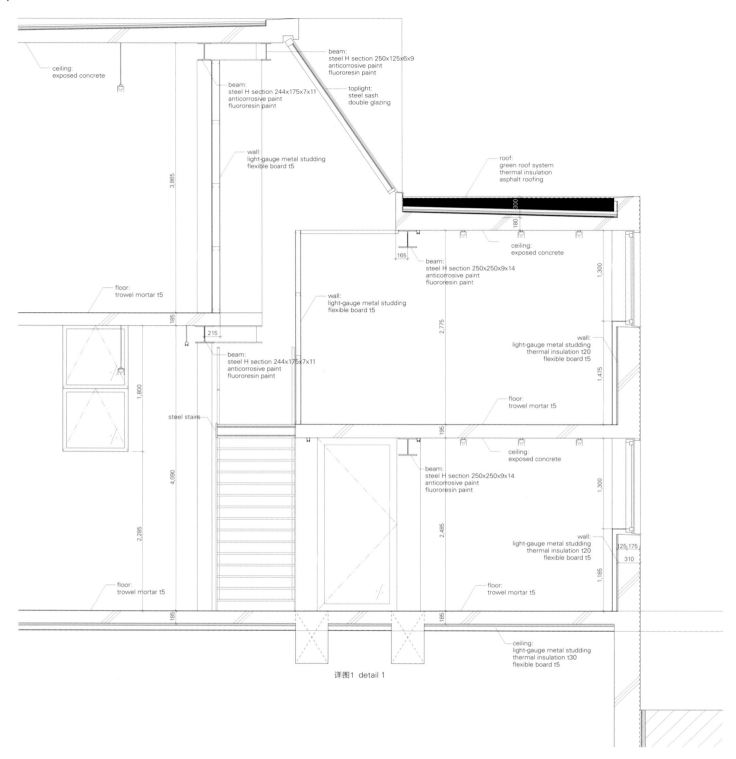

详图1 detail 1

## 隐藏的住宅
Teatum+Teatum

隐藏的住宅夹在两栋既有建筑物之间，远离街道，漠视城市的喧嚣，住宅内向而私密，与其所处位置相应和。

隐藏的住宅共分两层，室内各结构围绕着一座7m高的采光井布局。居住区直通住宅中央的采光井，从而使卧室和居住区可以相叠相连，这样的建筑结构使不同的功能区相联系。高处的天窗使光线照射进卧室和中央采光井，从而将室内与室外连接起来。由于在室内看不到室外的风景，因此便增强了这种室内感，营造出一种私密性，并将设计重点放在了光线和材料方面。

住宅的后立面是一个闪闪发光的黑色表层，上面嵌有二氧化硅碳粒，如同一幅面具，外人无法窥视住宅内的景象。经过激光切割的钢门，开合好似蝴蝶的翅膀，反映出冰冷的窗户上雨珠的图样。室内，激光切割的图案使光线可以照射进一层的隐藏空间。隐藏的住宅构造独特，为城市在有限的土地上以低价建造更多的房屋提供了范例。

A-A' 剖面图 section A-A'

## Hidden House

Hidden House is formed between existing buildings. Hidden House makes an opportunity of its dislocation from the street. It turns its back to the city and responds to its location by creating an architecture that is internal and intimate.

The house is organized over two levels and structured around a 7-meter high internal lightwell. Living spaces interface across the central lightwell, allowing bedrooms and living areas to overlap and connect. This interface between spaces seeks the opportunity for programs to infect one another. The connection to the exterior is formed through high level skylights that bring top-light to bedrooms and the central lightwell. By removing external views the sense of interior is reinforced, creating intimacy and a focus on light and materiality.

The rear elevation, a black shining surface, embedded with silica carbide particles acts like a mask, engaging the viewer without expressing or revealing the space behind. The steel butterfly doors are laser cut to reflect the pattern of rain on a cold window. Internally, the laser cut pattern allows light to penetrate into the hidden spaces of the ground floor interior. Hidden House provides a way for the city to create more housing on existing sites providing unique spaces at low cost. Teatum+Teatum

项目名称：Hidden House
地点：London, W12, England
建筑师：Teatum+Teatum
结构工程师：Fluid Structures
甲方：Teatum+Teatum
竣工时间：2012
摄影师：©Lyndon Douglas(courtesy of the architect)

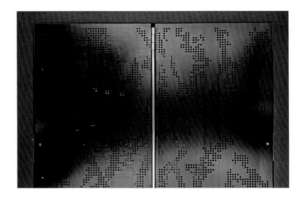

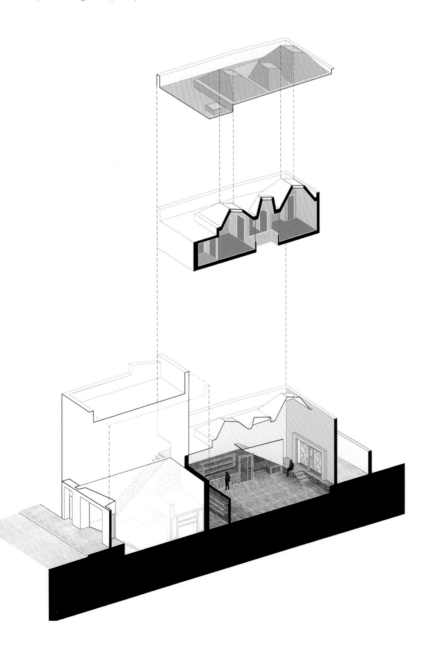

1 入口
2 卧室
3 采光井
4 浴室

1. entrance
2. bedroom
3. light well
4. bath room

二层 second floor

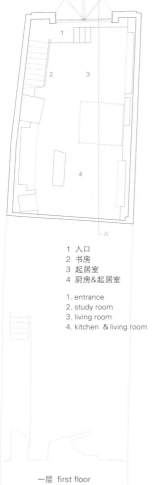

1 入口
2 书房
3 起居室
4 厨房&起居室

1. entrance
2. study room
3. living room
4. kitchen & living room

一层 first floor

# 阶梯住宅
Jaques Moussafir Architectes

该项目的基地上原来建有一座旧住宅，该住宅位于两座建筑（坐落在保存完好的法国巴黎第六区的中心地带）之间。该建筑被设计成一个树形结构，三面都被原有墙体围合起来。刷有白色涂料的原有石墙不禁让人想起当地根深蒂固的地方特色，而木质胶合板的树形结构又会让人心生现代之感。站在现有窗前可以俯视楼内的花园，现有的北墙和东墙根据历史古迹保护法保留下来，南墙则翻新为全玻璃式的，说明整座建筑几乎完全得到重建，同时透过亮晶晶的玻璃墙可以在不经意间瞥见室内空间的复杂性。

该项目最形象的比喻或许可称之为"巨石楼梯"，房屋如同一个楼梯，中心是防水淋浴间，楼梯间则是利用邻近建筑物的山墙建成的，楼梯和楼梯平台构成了起居空间。事实上，除了浴室，整个房间不需要分区，从地下室至屋顶露台没有间断，打造出一种空间连续感。屋子最大的特色在于其层叠的开放空间，而且从地下室到顶楼完全畅通：整个室内空间宛如一间150m²的房间，拥有八个不同高度的层面，除了天花板和地板，中间没有任何间断。树形的建筑物内只设有三扇门，分别通向地下室、地下室中的浴室以及顶楼的卫生间。

房屋整体是钢质结构，采用悬挑式地板，由基本独立于三面外墙的中心柱支撑，墙内镶嵌着混凝土盒形搁物架。建材的选择更凸显了建筑师的设计意图：中心柱的划分，地板、天花板都覆有洋槐木，它漆黑的颜色和花纹与外墙雪白的颜色和纹理形成了鲜明对比。

为了中和该住宅室内采用的漆黑面板，并使屋后洒满阳光，建筑师在房屋的东北角和西北角设置了两个向天空开放的采光井。屋内的交通流线呈竖直和螺旋状，人们在室内能够欣赏到周围这片保护完好的"拉丁区"建筑群中的花园和建筑。站在南立面的嵌壁式玻璃幕墙前，可以欣赏到朝向前院的主要景色。完全展开时，电控伸缩式百叶窗就起到了遮阳和保护隐私的作用，同时透过百叶窗还可以瞥见室内的复杂结构。金属屏风上激光切割的叶形雕花与环绕四周的绿叶相得益彰，点点树荫洒落屋内。

## Step House

Built on the site of an old house set between two buildings in the heart of a very well-preserved block in the 6th arrondissement of Paris, this house is designed as a tree-like structure delimited on three sides by the original walls. The whitewashed pre-existing masonry recalls the rootedness of the genius loci while the wooden veneered tree structure defines the contemporary spaces. The existing windows overlooking the gardens in the two remaining north and eastern walls have been kept according to historical preservation regulations whereas the entirely glazed south facade belies the almost total reconstruction of the building and provides a glimpse of the volumetric complexity of its interior spaces. Even more than that of a tree, the most effective metaphor for the project might be that of a Cyclopean stairway: the house is a stair whose core houses the wet rooms, whose stairwell is defined by the gables of the neighboring buildings, and whose steps and landings form the various living spaces. The fact that there was no need to partition the rooms (except the bathrooms) means that there is a sense of total spatial continuity from basement to roof terrace. The most interesting feature of the house lies indeed in its open space with stacked floors and no boundaries from the basement to the rooftop: the inside space appears like one single 150-square-meter room on eight different levels with no divisions apart from the ceilings and floors. Only three internal doors give access to the facilities located in the tree trunk, as for the cellar, the basement bathroom and the top floor toilet.

The structure, entirely of steel, is made up of cantilevered floors borne by the central core and partly dissociated from the three

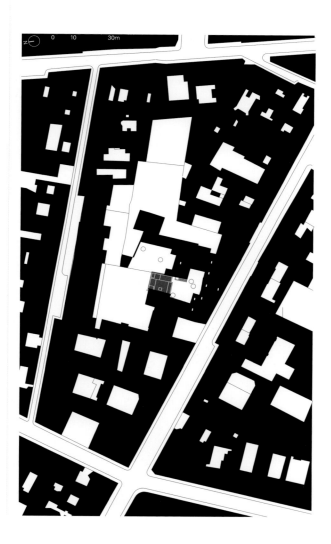

outer walls onto which have been grafted concrete boxes that act as built-in furniture. The choice of materials reinforces the architectural design: the partitions of the central core, the floors and the ceilings are all clad in locust tree, whose dark color and pattern contrast with the texture and whiteness of the outer walls.

In order to compensate the darkness of the veneer and bring light at the back of the house, two light-wells providing views towards the sky are located at the north-east and north-west corners of the house. The experience of the inside space is vertical and circular movement giving insights onto the surrounding gardens and buildings of this highly preserved cluster of the "latin quarter". The main inside views are orientated towards the front courtyard through recessed glazed curtain walls covering the entire southern facade. When unfolded, electrically operated shutters provide privacy and sunlight filter while still enabling glimpses on the volumetric complexity of the inside space. The laser-cut leaf shaped motifs of these metal screens blend with surrounding foliage and produce shades that paint shadows on the interior. Jacques Moussafir Architectes

项目名称：Step House
地点：22, Rue Jacob, 75006, Paris
建筑师：Jacques Moussafir Architectes
合作者：Alexis Duquennoy, Na An
结构工程师：Malishev Wilson Engineers
甲方：Eric de Rugy
总建筑面积：153m², 屋顶露台_15m²
造价：EUR 850,000
施工时间：2008—2011
竣工时间：2011.12
摄影师：©Hervé Abbadie (courtesy of the architect)

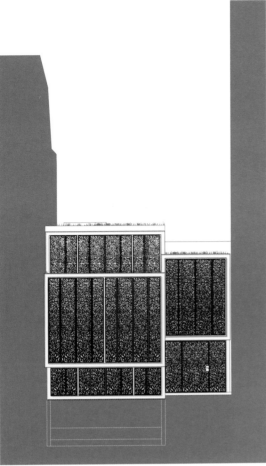

西南立面 south-west elevation

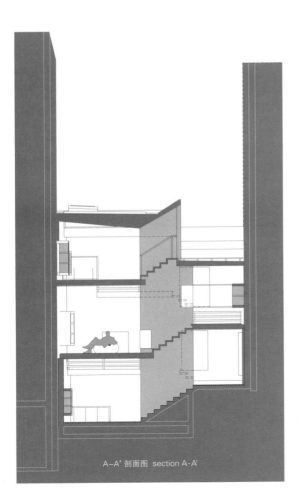

A-A' 剖面图 section A-A'

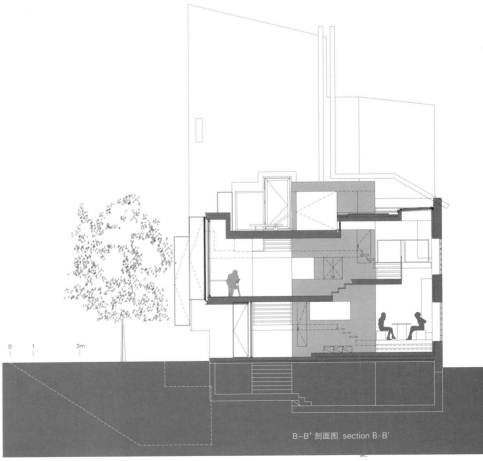

B-B' 剖面图 section B-B'

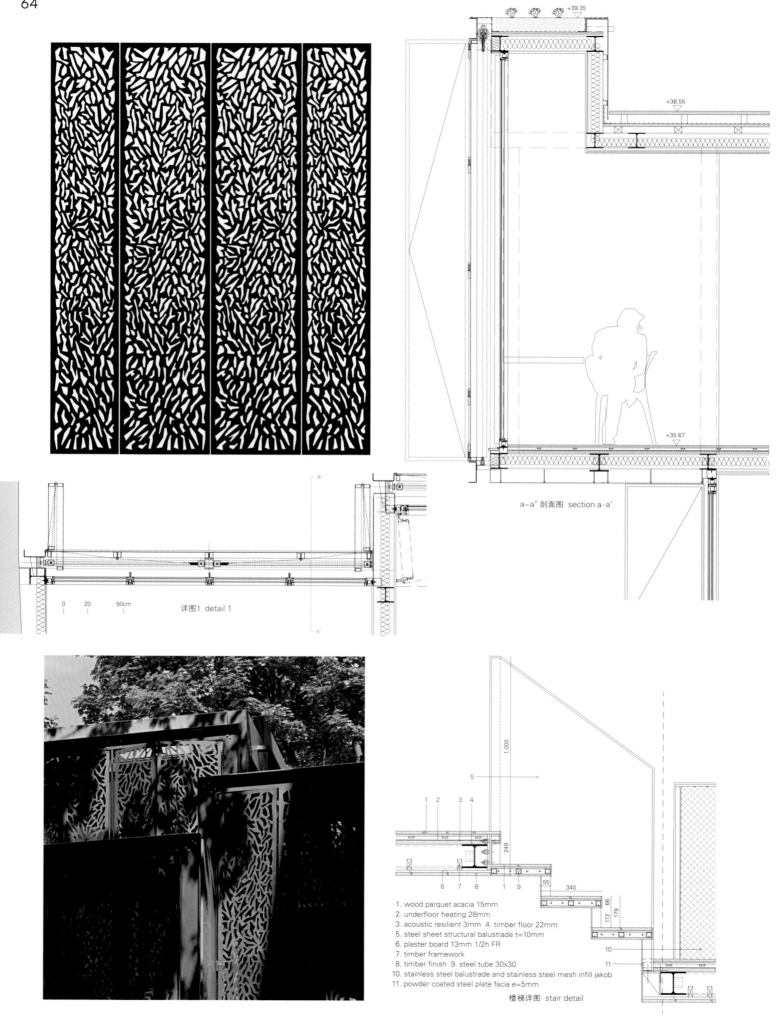

详图1 detail 1

a-a' 剖面图 section a-a'

1. wood parquet acacia 15mm
2. underfloor heating 28mm
3. acoustic resilient 3mm  4. timber floor 22mm
5. steel sheet structural balustrade t=10mm
6. plaster board 13mm 1/2h FR
7. timber framework
8. timber finish  9. steel tube 30x30
10. stainless steel balustrade and stainless steel mesh infill jakob
11. powder coated steel plate facia e=5mm

楼梯详图 stair detail

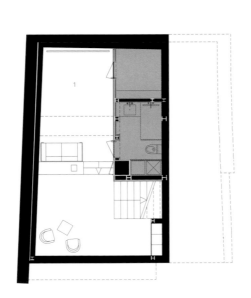

1 家庭影院

1. home cinema

地下一层  first floor below ground

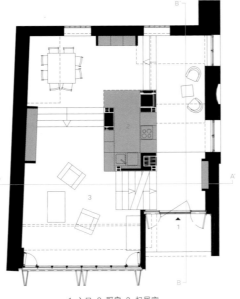

1 入口 2 厨房 3 起居室

1. entrance 2. kitchen 3. living room

一层  first floor

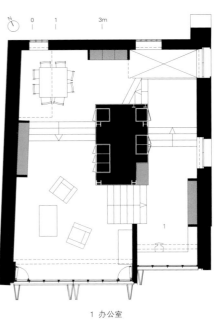

1 办公室

1. office

二层  second floor

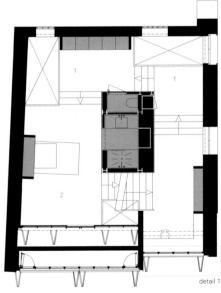

1 更衣室 2 卧室

1. dressing room  2. bedroom

三层 third floor

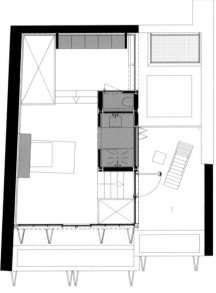

1 屋顶露台

1. roof terrace

四层 fourth floor

## 狄维娜护理之家14

Estudio de Arquitectura Javier Terrados

西班牙塞维利亚市人口稠密,街道、长廊、露台的设计都保留着古摩尔风格,狄维娜护理之家所在街道就位于该市的历史中心。该项目基地位于建筑群包围之中,只有一个6m高的开口面向街道,四围均是界墙。由于进入工地困难重重,土质松软,该项目迟迟未动工。土质松软的问题不容小觑,打桩的普通机械无法开入场地,因此就需要将软土挖出,填入实土夯实,才能打地基。

塞维利亚市政府计划为老年人和青年人提供一组面积为45m²的房屋,带有地下停车库,每套住宅有1~2个卧室,护理之家的中心为一个庭院,四周环绕开放式长廊,房间的房门面对长廊。长廊有一部电梯和两处楼梯,建筑周围有若干露台为卧室提供光照和通风。

庭院旨在通过建筑前厅成为街道的延伸,建筑师尽可能地将这个介于中间的前厅设计成一个透明结构。塞维利亚的传统建筑风格是在街道和露台之间设置过渡空间以保护住户隐私,同时还能欣赏到外面的美景。入口畅通无阻,既可行车也可步行,位于建筑立面的下部,建筑师将其设计成木质格架结构,作为半透明的隔膜。当地的一位涂鸦艺术家在墙底部画了一幅"抽象的树"的作品。从某种程度上来说,其深层含义远远不限于仅给人一种透明的感觉。

进入该建筑后遇到的第一个楼梯就位于前厅和庭院之间的角落,邻居们一出门就在前厅打招呼问候。每户都配有一个独立的邮筒,后面还设有一个软木公告板,公告板周围也留有一些空间。

庭院实质上就是传统意义上的公共露台,是居民交往活动的场所,庭院的设计初衷就是为居民提供相互交流的场地。庭院四周环绕网孔金属扶手,使建筑呈现出透明质感,露台远端的楼梯轻盈明快,镀锌管格架覆盖在折叠钢板之上,庭院上方有折叠遮阳篷,夏季可以享受清凉,远离安达卢西亚的炙热阳光。

该公寓将会用于出租,因此建筑师选用了镀锌钢、紧实的层压板、水磨石瓷砖、人造石等建材,以使建筑维护工作简便易行。

### Divina Enfermera 14

Divina Enfermera Street is located inside the historical center of Seville, a very dense urban tissue that keeps its ancient Moorish roots in the layout of its streets, passages and patios. The plot in which we were commissioned to build is an almost completely interior one, surrounded by party walls and with only a 6-meter opening to the street. It remained unbuilt, due to its access difficulties and because its soil of soft landfills. This last one was not a minor issue: the standard machinery for pile foundation could not enter the street. It was then necessary to substitute some layers of the soil with a more compact one as a base for a regular ground slab.

The Regional Government promoted a group of 45-square-meter dwellings for seniors and young adults, with a parking garage underground, the program called for one and two bedroom-units. The general layout used a single central courtyard surrounded by open corridors to which living rooms would open. Two staircases and an elevator serve these corridors. A series of small patios located at the perimeter of the plot give light and ventilation to the bedrooms.

That courtyard is intended to be the prolongation of the street through the building vestibule, an in-between space designed as transparent as possible. This kind of intermediate space between the street and the patio, where the privacy is guaranteed but the sight can flow from the outside is typical of Seville's urban tradition. Coherently, the continuous entry, for both pedestrians and cars, that forms the lower section of the facade is designed as a semi-transparent membrane, made of a lattice of wooden posts. A local graffiti artist collaborated with a painting of an "abstract

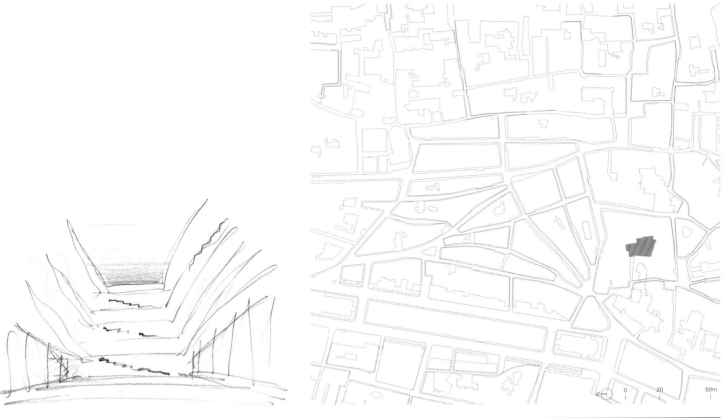

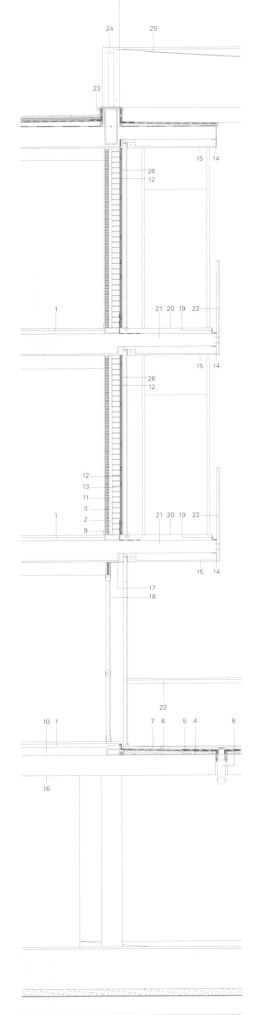

详图1 detail 1

1. 40x40cm terrazzo tile flooring
2. gypsum plaster on the wall and ceilings
3. 5cm hollow brick partition
4. oxiasphalt vapor barrier
5. expanded clay sloping
6. waterproofing asphalt membrane
7. 40x80cm artificial stone flooring
8. linear stainless steel drain
9. precast terrazzo base
10. 12cm expanded clay filling
11. 4cm air gap
12. 3cm sprayed polyurethane insulation
13. 12cm brick wall
14. folded galvanizaed steel drip
15. gypsum board suspended on the ceiling
16. silicate painting
17. folded galvanized steel sheet for lintel and jamb
18. casement/awninig window wirh aluminium sash and insulating 6+4+6 glass
19. 80.80.6 steel angle
20. 60x30cm ceramic tile flooring
21. reinforced concrete slab
22. folded perforated 5 mm galvanizaed steel plate railing
23. 1mm folded galvanized steel plate for covering over waterproofing membrane
24. railing formed with tubular galvanized frames and wire cloth panels
25. folding awning over patio
26. high-pressure compact laminate board fixed on galvanized steel battens

group of trees" on the wall at the bottom to somehow extend this sense of transparency and depth beyond.

The first of the two staircases is located in the angle between the vestibule and the courtyard. The vestibule is intended to be the first meeting place for neighbors. A free standing case for the mailboxes with a cork noticeboard at its back leaves some space around for that.

The main courtyard is a re-interpretation of the traditional corral de vecinos (communal patio), the scenery of social relations between neighbors. A general sense of transparency is translated also to the design of elements like the perforated metal hand railings that surrounds the courtyard. Also the staircase located at the patio's far end is very light, made of a simple folded steel plate covered by a lattice of galvanized tubes. A folding awning floating on top of the courtyard provides a ventilated shadow in the hot Andalusian summer.

The apartments will be for rent. So the use of galvanized steel, compact laminate boards, terrazzo tiles and artificial stone is intended to make easy the maintenance of the building.

Estudio de Arquitectura Javier Terrados

项目名称：Divina Enfermera 14
地点：C/ Divina Enfermera 14. Sevilla.
建筑师：Estudio de Arquitectura Javier Terrados
合作者：Marisa Jiménez León,
Rodrigo Morillo-Velarde Santos, Margarita Calero Santiago
结构工程师：TZ Ingeniería
质量检测员：Victor Baztán Cascales
施工单位：Fherlop
甲方：Empresa Pública de Suelo de Andalucía
用地面积：665m²
建筑面积：503m²
总建筑面积：2,138m²
竣工时间：2012
摄影师：©Fernando Alda(except as noted)

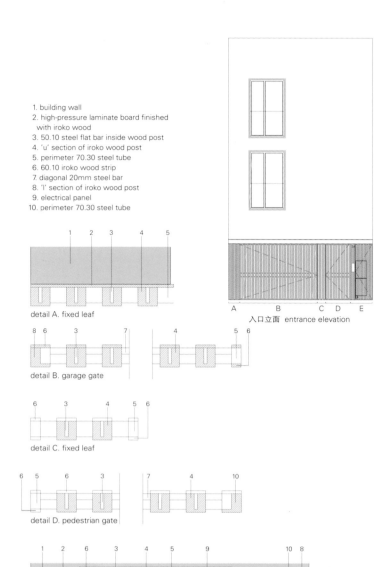

1. building wall
2. high-pressure laminate board finished with iroko wood
3. 50.10 steel flat bar inside wood post
4. 'u' section of iroko wood post
5. perimeter 70.30 steel tube
6. 60.10 iroko wood strip
7. diagonal 20mm steel bar
8. 'I' section of iroko wood post
9. electrical panel
10. perimeter 70.30 steel tube

入口立面 entrance elevation

detail A. fixed leaf
detail B. garage gate
detail C. fixed leaf
detail D. pedestrian gate
detail E. services entrance panel

A-A' 剖面图 section A-A'

B-B' 剖面图 section B-B'

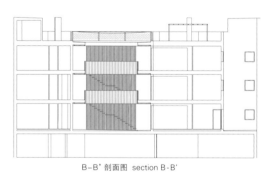

C-C' 剖面图 section C-C'

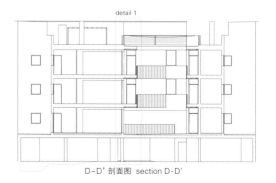

D-D' 剖面图 section D-D'

一层 first floor  二层 second floor

# 巴洛克庭院公寓
OFIS Arhitekti

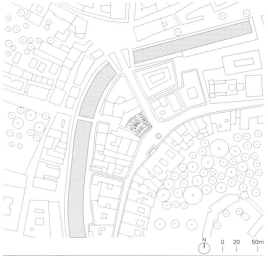

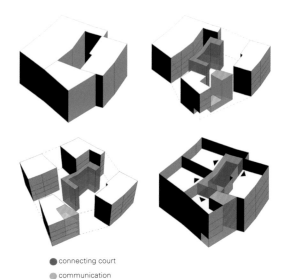

● connecting court
● communication

该项目包括改造三栋巴洛克风格的房子，附带一个封闭的内部庭院，巴洛克公寓位于老城区中心市政厅旁，罗巴喷泉对面，临近普莱克尼克的三座桥。这三栋房子都隶属于出版社，一层是书店，二层以上是办公室，该建筑的上一次改造还是在20世纪80年代初。这一次内部庭院将被改造成一个封闭的、半玻璃式服务空间，用来安置该建筑的主要空调装置。

项目提案要求将这些住宅连接成一个单一结构，同时将12座公寓围绕在小型内部庭院周围。巴洛克风格的临街面修复如初，而内部庭院则整修一新；这两项工作都在国家遗产部门的监管下完成。整修现存庭院的理念是将其作为楼层和公寓之间新的中央社交空间，同时为四周的公寓提供自然光照和自然通风，并且在公寓中可以俯瞰到庭院。由于庭院十分狭窄，且四面完全封闭，建筑师面临的主要问题就是如何使庭院拥有最佳采光条件，成为一个室内花园。最后所采取的方法就是在庭院立面上的历史构件上覆盖可以反光的玻璃。

在施工过程中，建筑师还发现了历史遗留下来的石拱、柱子和石板，并将它们应用在室内设计中。它们会倒映在玻璃外围护结构中。公寓空间的透明性以及反光构件，使庭院可以获得额外的光照，同时外部的结构也可以增加庭院的视觉效果，例如卢布尔雅那带拱顶的塔楼以及城堡都倒映在这个庭院之中。

## Baroque Court Apartments

This project involves the renovation of three Baroque block houses with a shared internal court in the old city center next to the city hall, opposite the Robba Fountain and close to Plecnik's three bridges. All of these three buildings are owned by a publishing company. The ground floor was used as a book shop and the spaces above on the first floor were used for offices and last adapted in the early eighties. The internal court was rebuilt as a

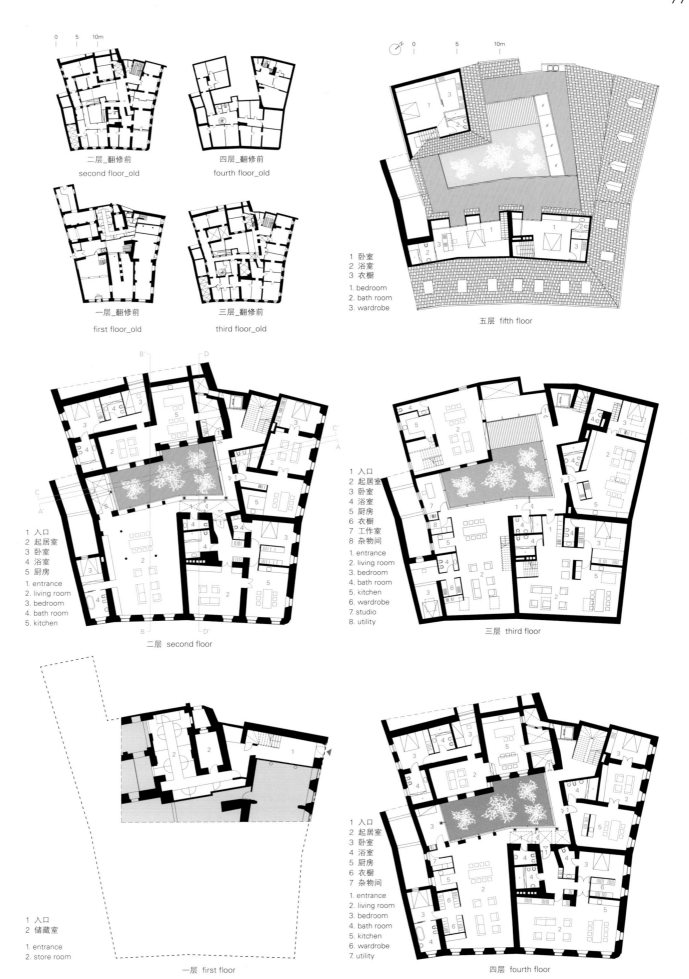

closed, semi glazed service space used for the building's main air conditioning devices.

The brief required to connect the houses into one single unit with the 12 apartments around the small internal court. Baroque elevations facing the street had to be mainly reconstructed to their original state. The internal court could be adapted and revitalized, but also under State Heritage's supervision. The concept reinstates the existing court as a new central communicating space between levels and apartments, bringing natural light and natural ventilation and cooling into the apartment spaces that overlook the court. Since the court is very narrow and enclosed from all sides, the main concern was to provide as much light as possible to become a form of internal garden. The result is a fully glazed and reflective envelope covering historic elements on the courtyard elevations.

Historical stone arches, pillars and slabs were discovered during the construction, and these elements become part of the interior. They form visible reflections on the glazed envelope. With the transparency of the apartment spaces and the reflected elements, the court benefits from additional light and visual elements from external features such as the tower of the Ljubljana Dome and the castle as they are reflected down into the courtyard. OFIS Arhitekti

A-A' 剖面图 section A-A'

B-B' 剖面图 section B-B'

C-C' 剖面图 section C-C'

D-D' 剖面图 section D-D'

1. floor construction:
   - terrace floor covering-teak 2.5+2.5cm
   - hydro isolation 2x
   - hibond slab
   - foil
   - steel beam
   - underconstruction
   - plasterboard
2. aluminium sheet, graphic color 2mm
3. aluminium spacer
4. palsterboard

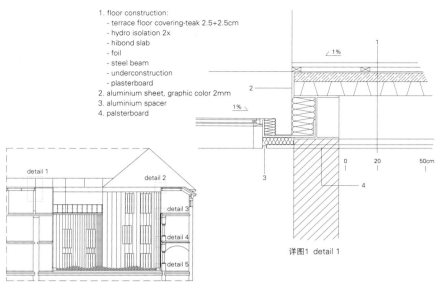

详图1 detail 1

1. terrace floor covering-teak 2.5+2.5cm
2. plasterboard
3. aluminium spacer
4. graphite color profile

详图2 detail 2

1. aluminium spacer
2. convector
3. graphite color profile
4. floor construction:
   - parquet 1.5cm
   - cement screed 5.5cm
   - separating layer-pe foil
   - thermal isolation 5.5cm
   - concrete 18cm
   - plaster 0.5cm

详图3 detail 3

1. aluminium spacer
2. convector
3. graphite color profile
4. floor construction:
   - parquet 1.5cm
   - cement screed 5.5cm
   - separating layer-pe foil
   - thermal isolation 5.5cm
   - concrete 8cm
   - plaster 0.5cm
5. fabric sunblind

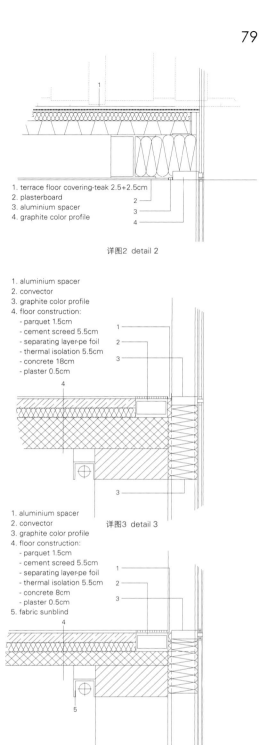

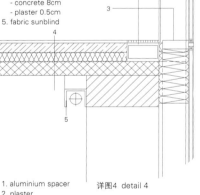

详图4 detail 4

1. aluminium spacer
2. plaster
3. graphite color mask
4. timber finish, Oak
5. sheet sink, graphite color
6. convector H=30cm

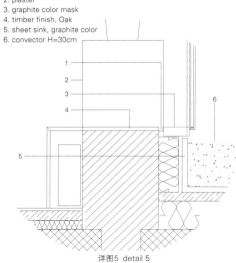

详图5 detail 5

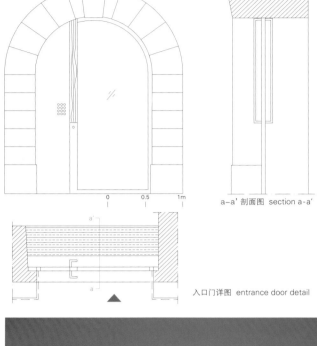

a-a' 剖面图 section a-a'

入口门详图 entrance door detail

项目名称：Baroque Court Apartments
地点：Ljubljana historical city center, Slovenia
建筑师：OFIS Arhitekti
项目团队：Rok Oman, Spela Videcnik, Andrej Gregoric, Janez Martincic, Janja Del Linz, Laura Carroll, Erin Durno, Leonor Coutinho, Maria Trnovska, Jolien Maes, Sergio Silva Santos, Grzegorz Ostrowski, Javier Carrera, Magdalena Lacka, Estefania Lopez Tornay, Nika Zufic
结构工程师：Elea IC d.o.o
机械工程师：ISP d.o.o
电气工程师：Eurolux d.o.o
项目规划：12 residential apartments, courtyard
建筑面积：692m²
内部空间：2,178m²
总建筑面积：2,420m²
造价：EUR 25,000,000
竣工时间：2012
摄影师：©Tomaz Gregoric & Jan Celeda

入口大厅详图 entrance hall detail

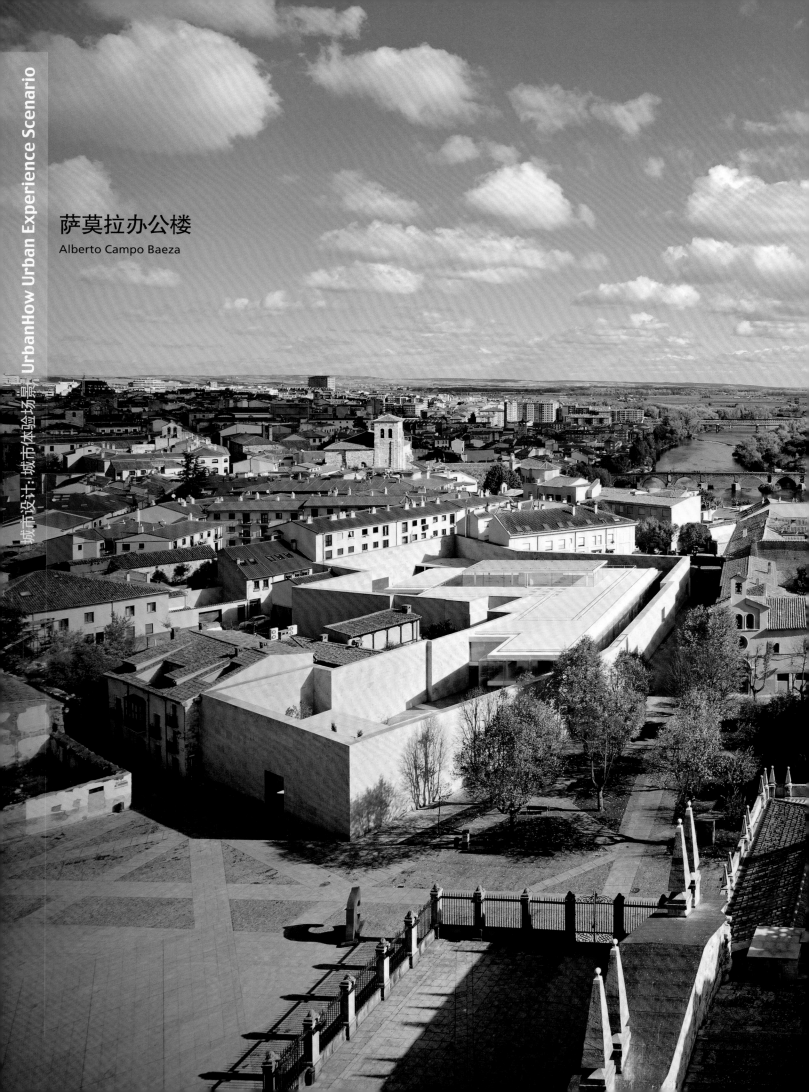

# 萨莫拉办公楼
Alberto Campo Baeza

### 与空气交融的建筑

修建一座与空气交融的耐久建筑，算得上是每一位建筑师的梦想。该项目的建筑师沿着老修道院的果蔬园的边界建造了一座面对着教堂的、用石墙围起来的盒状结构，朝天空开放。该结构的墙体和地面均采用石材，这种石材和教堂所用一致，宛若一座封闭花园。因此，建筑师在楼内修建了一处花园，花园内绿树成荫、花草芬芳。他们还在石墙上设置了几个开口，让内部的人们也能一窥墙外的大教堂、秀丽的风景和周围的建筑。建筑师对石墙的质量和规模进行了研究，以便使其能够展现出石头的坚韧，正如其在大教堂中表现得那样。相同的石材，但规模更广，厚度更大，这就凸显了该项目提案的优势。

面对教堂的一角，有一块250cm×150cm×50cm的巨石基座，上面凿刻着"HIC LAPIS ANGULARIS MAIO MMXII POSITO"的字样，代表着深植于泥土中的过去，即历史（译者注）。

在花园内部，建筑师修建了一个透明的玻璃屋，使人们感觉自己仿佛在花园中工作。在石质盒状结构中只有一个完全由玻璃打造的小屋，好似温室一样。它具有类似于特朗博墙的双层立面，建筑系统极其简单。立面可随气候的变化发挥调温作用，具有良好的保温效果，冬暖（温室效应）夏凉（通风立面）。立面外表完全由玻璃构成，每一块玻璃的尺寸是600cm×300cm×12cm，它们之间仅仅用一些硅胶连接，几乎看不到任何固定构件，看上去整座建筑构件仿佛是由空气连接在一起的。

这个盒状结构上部的三面体完全由玻璃制成，因此更加凸显了其透明性。这正是密斯希望在其费里德里希大街塔楼中所追求的效果。这个位于角落的三面体结构与空气融为一体，是一个名副其实的玻璃结构，上面刻着：HOC VITRUM ANGULARIS MAIO MMXII POSITO，意思是"未来"（译者注）。

石墙为记忆所建，地基深深扎根于土壤。
玻璃屋为未来而建，玻璃的棱角指向遥远的天际。
修建一座与空气交融的耐久建筑，是每一位建筑师的梦想。

## Zamora Offices

### Building with Air

To build with air, the abiding dream of every architect: facing the cathedral and following the outline of the former convent's kitchen garden, the architects erect a strong stone wall box open to the sky. Its walls and floors entirely made of stone, the very same stone as the cathedral. It's like a real Hortus Conclusus. They thus achieve a secret garden in which they conserve and plant leafy trees, aromatic plants and flowers. And they make some openings in these stone walls that frame, from within, the cathedral, the landscape and the surrounding buildings. For the stone wall,

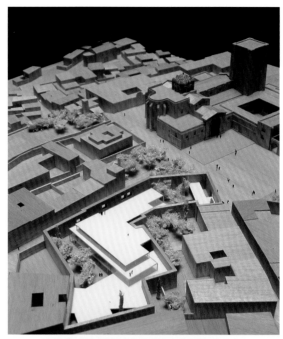

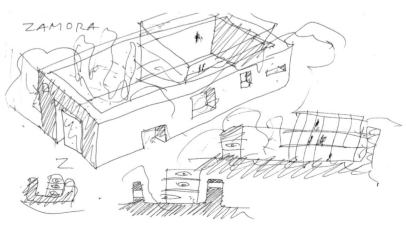

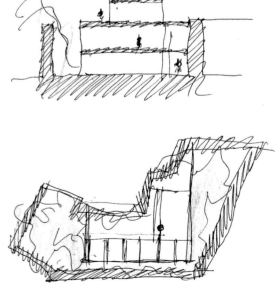

qualities and dimensions are studied to express the strength of the stone in the same way as it is in the cathedral. The same stone in large dimensions and with great thickness that accentuate the strength of the proposal.

In the corner facing the cathedral, a massive stone measuring 250 x 150 x 50cm, is a veritable cornerstone, and chiselled on that stone: HIC LAPIS ANGULARIS MAIO MMXII POSITO.

And in this garden, architects build a transparent glass box that makes it seem as if one is working within the garden. Within the stone box, there is a glass box, only glass, like a greenhouse. It has a double facade similar to a trombe wall, with maximum simplicity in its construction system. The facade works actively in regard to the climate, able to hold in heat in the winter (greenhouse effect) and at the same time to expel the heat and protect the building in the summer (ventilated facade). The external skin of the facade is made of glass, each single sheet measuring 600x300x12cm and all joined together simply with structural silicone and hardly anything else, as if entirely made of air.

The trihedral upper angles of the box are made completely with glass, thus even further accentuating the effect of transparency. Precisely what Mies was looking for in his Friedrichstrasse tower. The trihedron built with air, a true glass corner, and engraved in acid on the glass: HOC VITRUM ANGULARIS MAIO MMXII POSITO.

The stone box made from memory. With its cornerstone deeply rooted in the soil.

The glass box made for the future. With its glass corner blending into the sky.

To build with air, the abiding dream of every architect.

Alberto Campo Baeza

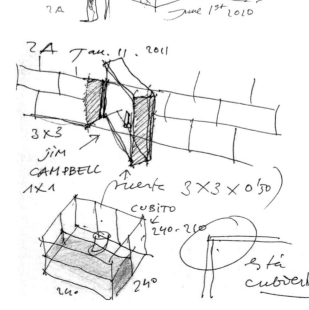

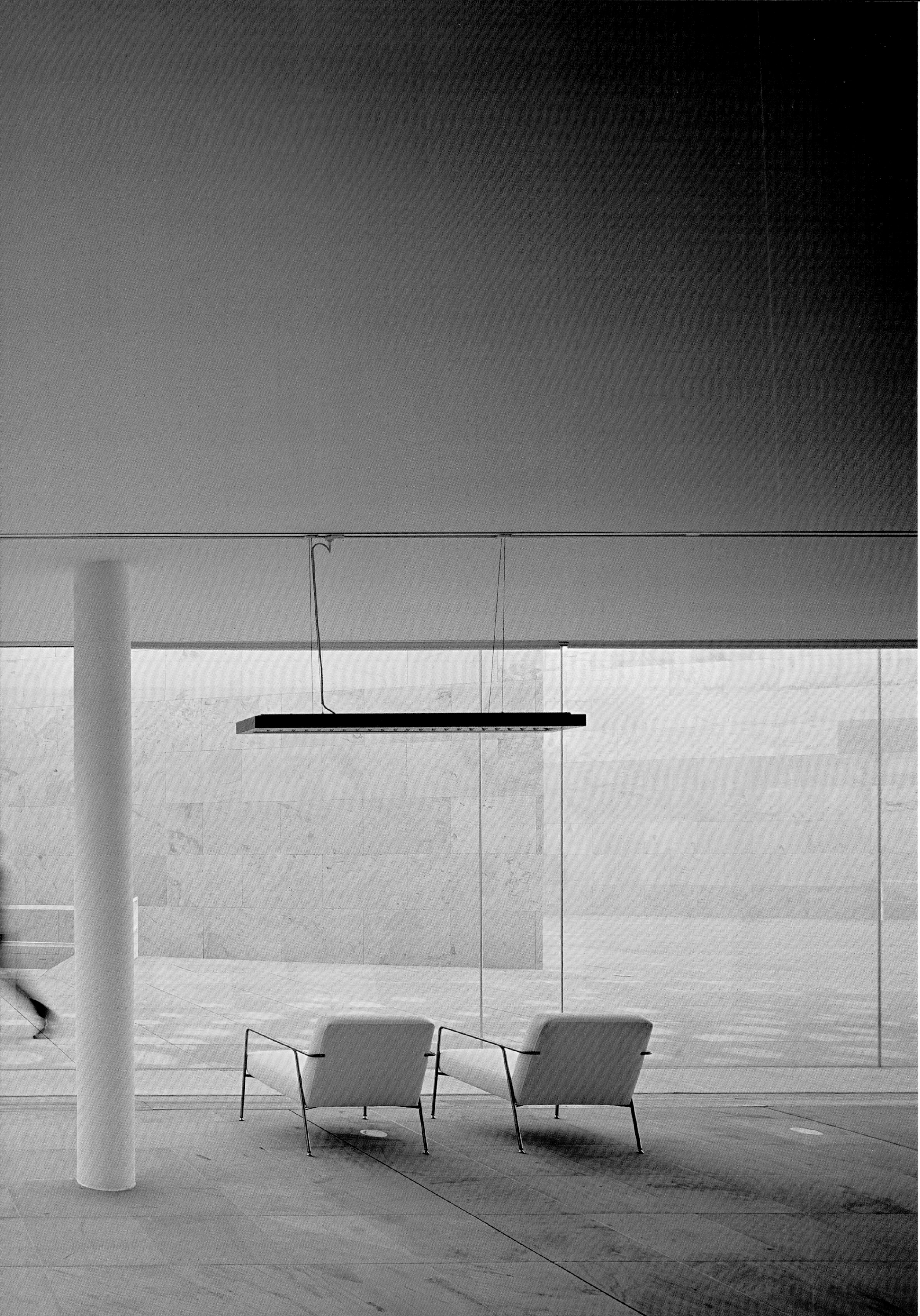

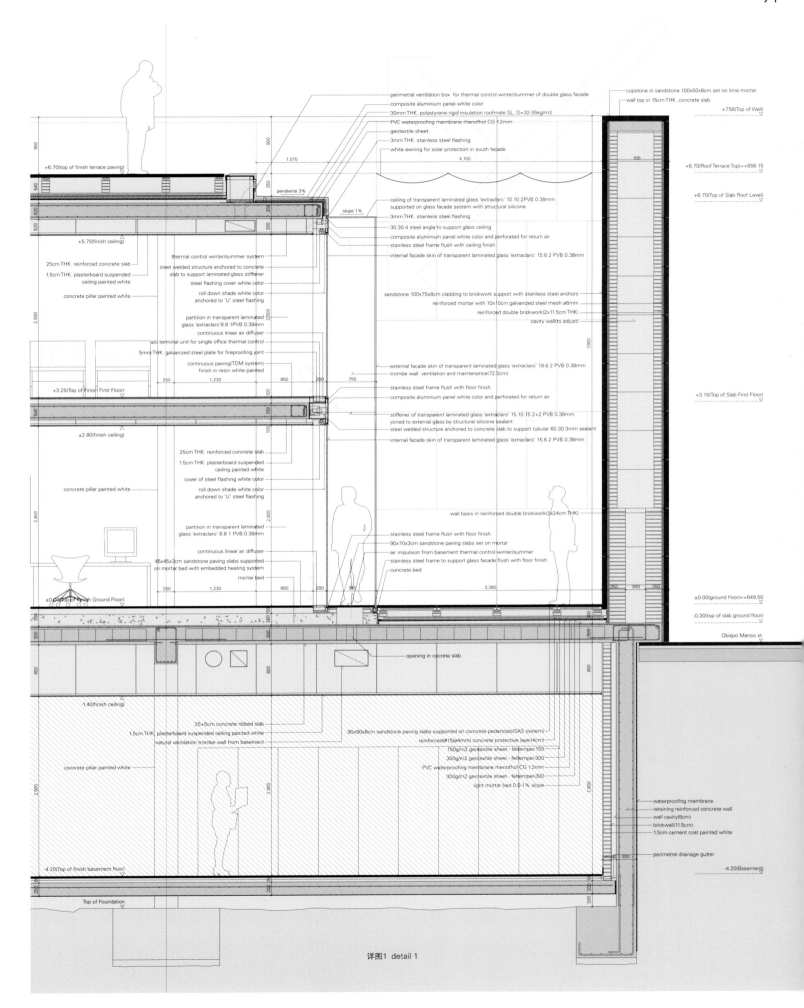

详图1 detail 1

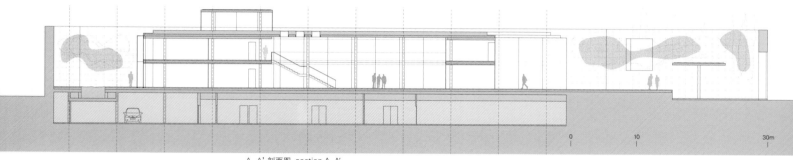

A-A' 剖面图 section A-A'

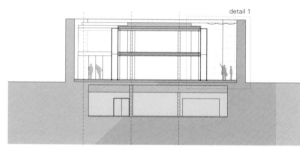

B-B' 剖面图 section B-B'

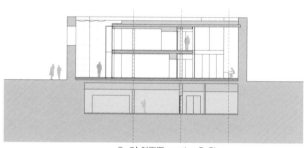

C-C' 剖面图 section C-C'

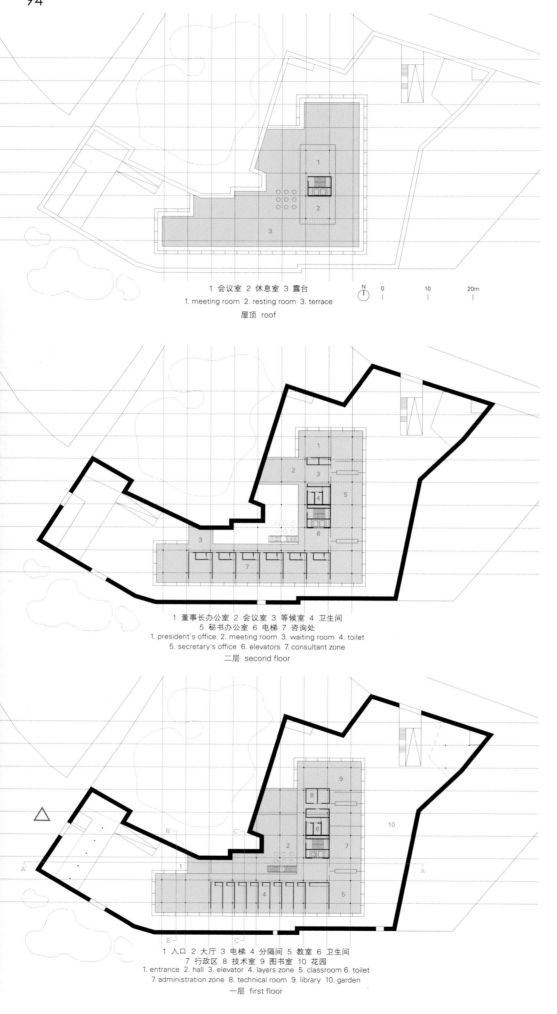

1 会议室 2 休息室 3 露台
1. meeting room 2. resting room 3. terrace
屋顶 roof

1 董事长办公室 2 会议室 3 等候室 4 卫生间
5 秘书办公室 6 电梯 7 咨询处
1. president's office 2. meeting room 3. waiting room 4. toilet
5. secretary's office 6. elevators 7. consultant zone
二层 second floor

1 入口 2 大厅 3 电梯 4 分隔间 5 教室 6 卫生间
7 行政区 8 技术室 9 图书室 10 花园
1. entrance 2. hall 3. elevator 4. layers zone 5. classroom 6. toilet
7. administration zone 8. technical room 9. library 10. garden
一层 first floor

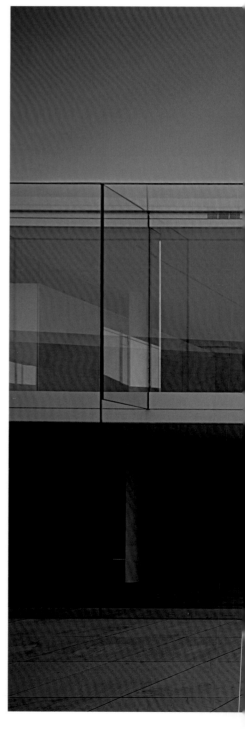

项目名称：Zamora Offices
地点：Obispo Manso, 1. Zamora
建筑师：Alberto Campo Baeza, Pablo Fernández Lorenzo, Pablo Redondo Díez, Alfonso González Gaisán, Francisco Blanco Velasco
合作者：Ignacio Aguirre López, Miguel Ciria Hernández, Alejandro Cervilla García, Emilio Delgado Martos, Petter Palander, Sergio Sánchez Muñoz
结构工程师：Eduardo Díez - IDEEE
机械工程师：Úrculo Ingenieros
照明设计师：Agosa
承包商：UTE Edificio Consejo Consultivo: Dragados - San Gregorio
检验员：Juan José Bueno Crespo
玻璃顾问：José Pablo Calvo Busello
甲方：Junta de Castilla y León. Consejería de Hacienda
面积：12,100m²
竣工时间：2012
摄影师：©Javier Callejas Sevilla (courtesy of the architect) (except as noted)

# 面朝街道，心系大海

# Facing the Street I am (not)

都市中的建筑物林林总总，相互争艳斗芳，只为能在众多的建筑物中脱颖而出，极具个性。本文要展示的项目反映了这一特点，并致力于获得全面的认可。整座城市成为一个建筑游乐场，身处其中的"玩家"各展身手，意在给游人一个惊喜，让他或她对这处空间或其外观印象深刻，以在这种建筑斗艳中被人们持续认可。

建筑的本质是人工系统的体现，这个系统与周围景观截然不同，甚至对景观的某些特点加以改动，以更好地适应居住其间的人们。希腊寺庙、哥特式大教堂和美国的摩天大楼都是由视觉分隔、独特的原创性、装饰和结构方面的独具匠心以及其本身的独树一帜特点构成的空间元素。然而，这些建筑范例的作用非同一般，它们把独特性和功能性规划的图解结合在一起。

The projects presented here reflect the need to stand out, to emerge, in an urban territory made of a set of architectural works that compete, often vehemently, to bring attention to their own personalities, as the projects strive to achieve full recognizability. The city becomes a kind of architectural playing field, on which each player tries to surprise the visitor, impressing upon him or her the memory of this space, this shell, in the duel for lasting recognition.
Architecture is, in its essence, the emergence of an artificial system which differs clearly from the surrounding landscape, changing its characteristics in order to accommodate the people living inside it. Greek temples, Gothic cathedrals, American skyscrapers are spatial elements that make of their visual separation, their peculiar originality, their decorative and structural effort, and the distinctive feature of their being. They are examples, however, of architecture with exceptional

# a Duck

接下来我们将要讨论的项目强调未来的发展趋势,即采用"普通"的功能来将建筑转为具有毫无限制的原始性的项目,使其从城市构造中脱颖而出,获得相应的历史感。

简而言之,围绕建筑物自身与其周边的环境的联系,我们所做的努力可以归结为两种方法:一类是明晰法,建筑可以采用这种方法来进行开发,好像大街上突然有人高声喊叫一般;另一种是较为含蓄地嵌入,倾向于激起观看者的好奇心。不论何种办法,它都可能有助于我们通过诠释这些形式,来将建筑联想为一部电影、一本书和一处示范性建成空间。

purposes, linking their peculiarity to the iconography of the functional program.
The projects we discuss below, however, emphasize the further trend of transforming architecture with "ordinary" functions into events of unbridled originality, in an attempt, to have them emerge from the urban fabric of the city, to become more or less historical.
These attempts can be reduced to two approaches which link their stories to the relationship with the surroundings: they may be developed via an explicit method, as a kind of scream into the street, or via a more veiled intervention, which tends to entice the curiosity of the viewer.
In either case, it may help us in explaining of these modes to relate the work to a film, a book, an exemplary built space.

27D大楼/Kraus Schönberg Architects
墨西哥城西班牙文化中心/JSª Arquitectura + Arquitectura 911SC
梦想商业中心酒店/Handel Architects
巴黎篮子学生公寓/OFIS Arhitekti

面朝街道，心系大海/Diego Terna

House 27D/Kraus Schönberg Architects
Spain's Cultural Center in Mexico/JSª Arquitectura + Arquitectura 911SC
Dream Downtown Hotel/Handel Architects
Basket Apartments in Paris/OFIS Arhitekti

Facing the Street, I am (not) a Duck/Diego Terna

### 我即丰碑

1995年，Jim Jarmusch推出了名为《死人》的电影，借助主人公一生中发生的一系列转瞬即逝的小插曲，利用一个简短的故事，讲述了一段长长的旅途，这些插曲虽无足轻重，但是却透漏着冷漠的元素，让人感到不舒服。黑白的剪辑图像通过连续的间断和活动交替，向人们展示了主人公乘坐火车途经几千英里，横跨19世纪的整个美国所看见的交替变换的景观。

这场离奇的旅途充满悬念、回眸犀利但却十分狭窄、记忆徐徐展开，充满无限期许。电影给人的感觉是故事只是借助火车窗外开放空间的连续性空间变化（自然的或人工的）展开的，如同移动的图片一样，这些空间的出现和变化刻画出了一国之根本、铁路沿线展开的景观以及使景观熠熠生辉的纪念物。然而，Jarmusch借助一连串精心安排的静态场景与交流的景象联系起来。类似的，Robert Venturi、Denise Scott Brown和Steven Izenour于1972年在合著的《学习洛杉矶》一书中阐述了逆空间的设计策略，旨在采用赋予建筑的信息交流来建设。

Monticello汽车旅店的标志如同奇彭代尔式高橱的侧影，远远地在其前面的高速公路上就可以看得见。这种建筑风格的标志即是逆空间的，体现了空间之间的交流，交流在建筑和景观中是主导因素。但是它适用于设计新型景观。旧式折中主义的哲学联想所唤起的微小的和复杂的意义只能在传统景观较易改造的空间中细细品位了。但是商业的介入，折中主义对路边开阔且复杂的景观背景产生了巨大的影响，这些景观规模大，发展速度快，规划复杂。不同的建筑风格和标志把多种各有千秋，日新月异的因素联系起来。其商业气息浓厚；建筑环境几乎是全新的。

Venturi、Scott和Izenour三人在描述洛杉矶这一类非典型性城市的过程中，形成了20年后Jarmusch利用黑白镜头记录旅途上发生的连续事件所展示的建筑术语：一片疆土，城市或自然，建筑作为一种交流体系明确地存在着，使景观变得更亲切，好似一本打开的书，有待从整体上进行赏析，通俗易懂，即使非专业的读者也能读懂。这种设计和空间没有很大的关系，因为它与二维形象概念相联系，通过不断变换的框架得以体现，正如《死人》中通过车窗所看到的景色。

### I am a Monument

Jim Jarmusch introduces the 1995 movie *Dead Man* with a brief tale of a long journey, related through a series of fleeting sequences of a few moments of the protagonist's life – episodes apparently negligible, but which contain some alienating elements and thus discomfort us. The editing of the black and white images shows us, through a continuous movement of breaks and activities, the alternation of landscapes that the train, which carries the protagonist, travels along the thousands of miles necessary to cross the US territory of the 19th century.

And in this surreal journey made of suspended moods, of penetrating yet incomprehensible glances, of memories slow to emerge, of infinite expectations, it seems the story unfolds only through the continuous appearance of spatial emergences (natural or artificial) in the open spaces beyond the window of the train: as moving pictures, they can tell, by their presence and the continuous change, the essence of a nation, its unfolding along the railroad tracks, the monumental need for shimmering landscapes. Somehow Jarmusch relates a vision of communication made through static images that discreetly follow one another. Similarly, in 1972, Robert Venturi, Denise Scott Brown, and Steven Izenour in the book *Learning from Las Vegas*, write of an antispatial design strategy, constructed through the communication of a message given to architecture.

*The sign for the Motel Monticello, a silhouette of an enormous Chippendale highboy, is visible on the highway before the motel itself. This architecture of styles of signs is antispatial; it is an architecture of communication over space; communication dominates space as an element in the architecture and in the landscape. But it is for a new scale of landscape. The philosophical associations of the old eclecticism evoked subtle and complex meanings to be savored in the docile spaces of a traditional landscape. The commercial persuasion of roadside eclecticism provokes bold impact in the vast and complex setting of a new landscape of big spaces, high speeds, and complex programs. Styles and signs make connections among many elements, far apart and seen fast. The message is basely commercial; the context is basically new.*

Venturi, Scott and Izenour, in describing an atypical city like Las Vegas, simply frame in architectural terms what Jarmusch, twenty years later, will show through the sequences of a journey in black and white: a territory, urban or natural, in which the architecture emerges clearly as a communication system that, in this manner, makes the landscape more familiar, as a sort of open book to be viewed in its entirety, easily understandable even by a non-specialized spectator. It is a design that has little to do with space, since it is linked to the concept of the two-dimensional image, exemplified by the changing frames, viewed through windows, of *Dead Man*.

拉斯维加斯一家甜甜圈商店的标志
the sign for a donut shop in Las Vegas

位于拉斯维加斯的、Robert Venturi 称之为"装饰性小屋"的奇幻标志
fancy signs in Las Vegas that Robert Venturi called "Decorated Shed"

　　Neutelings Riedijk Architecten设计的于1997年开放的Minnaert大楼，可以使人们认可上述方法：大红色元素，覆有粗糙的石膏板，建筑表面有纹理。这里的每种建筑元素看起来都想通过一系列装饰设计（清楚地表达了想要给参观者惊喜的意愿），来传递出强调交流的理念。

　　例如入口处的门廊，没有采用显而易见的结构方法，而是创造性地使用立柱，在上面用字母拼写出大楼的名字。游客入内时必须经过这些似乎会说话的立柱，交流在这里变得如此清晰，其本身成为一种建筑形式。

　　大楼成为一种空间叙事声明，通过彰显的戏剧性行为吸引前来参观的游人，带领他们深入建筑物的内部，内部更加引人入胜：这里没有干扰，环境十分复杂，一处接着一处，从中央大厅（带有一个雨水蓄水池和容纳阿拉伯式工作室的壁龛）到穿孔天花板（使人想起布满星星的夜空），再到鲜亮的柱子，让人不禁想起怀特的作品。

　　荷兰建筑师们创造了富有特殊力量的建筑杰作，犹如一台风景如画的机器，完全吸取了Venturi的理念，借助建筑作品实现了与游客的直接且易于理解的交流。

　　Handel建筑师事务所设计的梦想商业中心酒店和位于巴黎的、由OFIS Arhitekti设计的篮子公寓都是一系列独特的事物组成的系统，这些独特的事物抵制夸张的做法，旨在获得全方位的可视性，并且成为市区的代表性建筑。这主要是通过非常规的建筑外形、丰富的装饰性元素来塑造的；其目的十分明确，即通过纯粹地利用建筑外形，来使建筑清晰可变，而不管其空间大小。酒店建筑的表皮犹如面纱，圆形洞口穿过其中，表皮固定在无法安装铰链的网格上，这一主题伴随着客人进入到每一处环境，客人目光所到之处都有融入环境的感觉。

　　篮子公寓采用了建筑的基本模块——房间来容纳学生，作为解决狭长地块的主要元素，为此，通过添加网格，建筑改变外形，其相同的模块落成螺旋形。

　　从Jarmusch的想象、Venturi的描述和Neutelings Riedijk的创意中人们可以注意到，这两个项目其实都是要寻求一种在城市中释放的途径，以宣告它们的存在。

**我是一尾大鱼**

　　我们只有在2003年由Tim Burton执导的影片《大鱼》的片尾才

It is possible to recognize this approach in the Minnaert building, designed by Neutelings Riedijk Architecten, opened in 1997: a great red element, covered with rough plaster, with "veins" that emerge from the surface of the building. Each architectural element here seems to convey an intentional emphasis on communication, through a series of decorative inventions that make explicit the desire to surprise the visitor.

The entrance portico, for example, abandoning any evidently structural approach, invents a kind of column style, using letters of the alphabet to compose the name of the building. It is through these speaking columns that visitors must pass to enter: Here, explicitly, the communication is so clear as to be a form of the architecture itself.

The building thus becomes the manifesto of a spatial narration that draws visitors through acts of overt theatricality and leads them into the bowels of an architectural body that never ceases to amaze even in the interiors: Here, without interruptions, environments of rich complexity follow one upon the next, from the central hall, with a reservoir of rainwater and arabesque studio niches, to perforated ceilings, reminiscent of starry skies, to bright columns which mention, by contrast, the work of Wright.

The Dutch architects build, therefore, a masterpiece of unusual force that works like a scenic machine, well aware of the lessons of Venturi, which tell of an architecture that is able to touch the visitor with a direct and understandable language.

The Dream Downtown Hotel by Handel Architects and the Basket Apartment in Paris by OFIS Arhitekti work on a system of continuous peculiarities that brush up against exaggeration, with the intent of gaining full visibility and becoming iconic in the urban surrounding. They do so through unusual shapes, full of decorative elements, with the clear aim of making the architecture explicit through a pure use of shapes, regardless of the space each builds. The hotel uses the skin of the building as a veil pierced through by circular openings, arranged on a grid impossible to unhinge, a leitmotif that accompanies the guest into each environment and informs one's interaction with all the spaces it overlooks.

The Basket Apartment uses the basic module of the building, a room for students, as the key element to solve a long and narrow site, therefore building by addition upon a grid deformed by the continuous rotation of the same modules.

Mindful of the visions of Jarmusch, the descriptions of Venturi, the inventions of Neutelings Riedijk, these two projects seek a way to unleash their scream on the city, to proclaim their presence.

**I am a Big fish**

Only at the end of the movie *Big Fish*, directed by Tim Burton in 2003, do we understand that the stories told by the protagonist to his son, so fantastic, hide, in truth, real characteristics, made clear only in death. These are stories that tell of an amazing humanity,

Neutelings Riedijk事务所于1997年在荷兰建造的Minnaert大楼
Minnaert building, the Netherlands, by Neutelings Riedijk Architecten, 1997

会懂得主人公讲给他儿子的故事,如此的难以置信、深藏不漏、富有哲理、人物真实,并且只能以死亡而告终。这些故事表达了人性的不可思议性,因富有活力、人物各异、举止和行为超乎想象,而让观者印象深刻。

在主人公葬礼上,我们注意到故事中的虚构的美好角色事实上是想象力重组的结果。这个重组给这些同样的人们一个神秘的光环,或者扭曲了他们最初的形体特征。

美国导演执导的这部电影向我们展示了两个并行不悖的宇宙的故事,一个宇宙包含在另一个之中,出现在不同的场景——在故事中是非现实的、童话般的,但在正式场景却变得如此现实且可信的。然而,这两个宇宙不分伯仲,因为所有人物都在葬礼上出现了,展现了一种局限性,人物也不是毫无拘束的,这使得他们分别存在于两个平行的宇宙中。

为了满足这种同时性,划清现实和幻想的界线是非常必要的,这条界线对应用领域进行区分,是虚拟世界和现实世界的分水岭。

1923年,勒·柯布西耶在《关于建筑》一书中确认了建筑的入门知识,即对整治线进行定义,把整治线背后的几何规则附加在一系列历史建筑物的立面上。

整治线对设计的反复无常性加以限制:这样方法可以确保所有作品创作源于热情,如同学生的九分法以及数学家的推理证明。整治线满足了精神层次的顺序,使人们追求灵巧和和谐的关系,它赋予了建筑物节奏感。

整治线引入了数学的条理性,有条不紊,让人舒心。它解决了建筑中基本的几何问题;也解决了建筑物的一个"基本特点"问题。整治线可谓带给建筑业决定性的灵感启示,是建筑中十分重要的操作步骤。

大楼叠加的部分显示了表面的重要性,与其说是像皮肤,不如说是在大楼内外之间、外部观者和楼内人们之间发挥协调作用的一层膜。大楼的立面是遵循黄金分割的原则建造的,成为一种与城市环境交流的手段,而这种交流并没有遵循Robert Venturi制定的规则:在该案例中,它遵循的是Le Corbusier从原始时期带回的、描述一种与生俱来的感觉的线条规则,这种规则并不是通过直白的语言而是借助无声的暗示,不知不觉中给人留下深刻印象来进行交流的。

able to impress the viewers with surprising vitality, with a variety of incredible characters, with actions and behaviors that seem to exceed even the possibilities of imagination.
Yet at the funeral of the protagonist, we observe that these wonderful fictional characters are, in fact, the result of an imaginative reworking which gives these same people a kind of mythical aura, exaggerating or distorting their primary physical characteristics.
The film by the American director thus tells the story of two parallel universes, one contained in the other, appearing in different situations – unreal and fairy-like in the tales, realistic and credible in official situations. However, neither has supremacy over the other, we understand, as these characters come together at the funeral, showing a kind of limit, a boundary of personality that allows them to be on one side and the other of the two parallel universes.
For this simultaneity to happen it is therefore necessary to define a border between reality and fantasy which distinguishes the application field, forming the watershed between the world of fiction and the world of reality.
In 1923, Le Corbusier, in the book *Towards an Architecture*, identifies a possible architectural threshold that defines Regulating Lines, transposing the geometric rule behind them onto the facade of a series of historic buildings.
*A regulating line is an assurance against capriciousness: it is a means of verification which can ratify all work created in a fervour, like the schoolboy's rule of nine, the Q.E.D. of the mathematician. The regulating line is a satisfaction of a spiritual order which leads to the pursuit of ingenious and harmonious relation. It confers on the work the quality of rhythm.*
*The regulating line brings in this tangible form of mathematics which gives the reassuring perception of order. The choice of regulating line fixes the fundamental geometry of the work; it fixes therefore one of the "fundamental characters". The choice of the regulating line is one of the decisive moments of inspiration, it is one of the vital operations of architecture.*
This work of overlap reveals the importance of the surface that surrounds the building, not as a skin, but rather as a membrane for mediation between the outside and the inside of the building, between the one who observes from the outside and the one who lives in the interior. The facade of a building, constructed geometrically under the rules of the golden section, becomes a means of communication with the urban environment, a communication that does not, however, follow the rules set out by Robert Venturi: in this case the rule of the Lines, which Le Corbusier brings back from primitive times and which describes a sort of innate feeling, communicates with people not through explicit language, but through a silent suggestion, which affects the unconscious.
In this sense, then, the facade enclosing a building allows one to

Herzog & De Meuron建筑事务所于2000年在法国巴黎建造的Rue des Suisses住宅楼
Rue des Suisses in Paris, France, by Herzog & De Meuron, 2000

朝街的西班牙文化中心的加外框的立面
a framed facade of Spain's Cultural Center facing the street

从这个意义上来讲,包围大楼的立面允许人们进入到另一个平行的世界,即大楼立面包裹下的封闭世界,如同神奇冒险片《大鱼》里描述的、受到保护的宝贵空间一样。

这个神秘的入口在Herzog&De Meuron建筑事务所于2000年在巴黎建成的Rue des Suisses住宅楼上得以实现。该立面有一系列深色可移动的金属板构成,在封闭的时候,它将大楼的前面改造成一个不为人知的匣子,使表面形成一个障碍,吸引外面的人们深入探索。它作为一种与其所在的城市互动的元素发挥作用——通过刚性的但却呈现友好态度的几何布局,这多亏了大楼前方内部的皱折——同时,它也是设有庭院的一个微观世界。

深入建筑内部可以探索到与外面世界并行的世界,即完美复原了外部的几何造型,但是蜿蜒曲折、轻质木包裹着住宅阳台的世界。

面向城市的深色立面似乎想要保守住内部的秘密,只对想方设法进入院内的少数人展示自己,等待他们来发现边界处一截低矮的、长长的灰色墙体和一些只有以时尚住宅的形式出现的元素。

那么到底什么才是建筑的真正形式呢?正如Tim Burton所言,也许它包罗万象:外城、威严的立面、宏伟的院子,就像电影里的主人公一样,在结尾的时候才发现可能是一尾大鱼。

所以由Kraus Schönberg建筑师事务所设计的27D住宅和JSa Arquitectura + Arquitectura 911SC 设计的西班牙文化中心中交替利用他们的故事,使外观极具可识别度,但是又不像前面的那些项目那么拖沓:建筑形式寻求一种牢固的几何形状,不外露,但是框架结构完全遵守严格的规则,且十分重视内部的空间,注重保持和保护组成室内的最敏感的内部环境。

27D住宅沿着灰色的立面严谨地布局了一些洞口,重新利用了周围大楼的历史性线条,并且蜿蜒穿过一座庭院,院内的大型露台布满阳光。

西班牙文化中心将主立面一分为二,两部分方法各异(根据其关系的亲密程度):沿街下行的部分巍然耸立着,非常空洞,好像在寻找与这个城市对话的力量;上行的部分视野开阔,立面移动以寻求阳光,是俯视庭院的大门,好像乐曲的前奏一般。

enter into a parallel world, the one enclosed in the building that is "behind" the facade, as if protecting a precious world similar to the one narrated in the amazing adventures of Big Fish.
This mysterious portal is achieved in a residential building by Herzog & De Meuron, completed in 2000 in Rue des Suisses in Paris. Here the facade is constituted of a series of movable panels of dark metal which transform the front, when closed, into an indecipherable casket, rendering the surface a barrier that attracts one to that which is beyond. It operates, therefore, as an element that can interact both with the city where it is located – though a geometry that is rigid, yet friendly, thanks to the folding of the front inward – and with the microcosm that includes the courtyard.
And it is in this interior that there seems to explode a parallel world, a wonderful revival of the external geometries, but here frizzy and sinuous, in light wood, which encloses the balconies of the houses.
It is as if the dark facade facing the city wants to keep the secret of the world that will be revealed only to the few who manage to enter the court, to discover a low body, a long grey wall on the border, some small elements in the form of a stylized house.
What, then, is the true form of this architecture? As told by Tim Burton, maybe it encompasses all: the outer city, the stately facade, the marvellous court, like the protagonist of the film, which in the end turns out to be, perhaps, a big fish.

Thus the House 27D by Kraus Schönberg Architects and the Spain's Cultural Center by JSa Arquitectura + Arquitectura 911SC rotate their stories on an external recognition that is well marked, but is less redundant than that of the previous projects: the forms of architecture seek a solid geometry, non-explicit, but well framed within strict rules, containing spaces with greater emphasis on the interiors that maintain and protect the most sensitive environments that in fact make up the interior spaces.
The House 27D has a strict arrangement of openings along the grey facade, thus resuming the historical lines of the buildings that surround it, winding through a courtyard that accommodates large terraces open to the light.
The Spain's Cultural Center divides its main facade into two parts, implemented in two distinct ways depending on the ratio of the relationship: downward, along the street, it stands solemn, with the full prevailing on the empty, as if looking for the force required to support the dialogue with the city. Upward, it opens onto views, moving the facade in search of the light, a prelude to the great outdoors overlooking the courtyard. Diego Terna

## 27D大楼
Kraus Schönberg Architects

城市设计：面朝街道，心系大海　UrbanHow Facing the Street, I am (not) a Duck

该项目需要考虑周围的历史性建筑及其30m长的地块,并将建筑物插入10m宽的临街部分。

这栋多用途大楼朝向商业街,南边的日光照进院内,南侧的拱廊在相对稠密的历史市中心提供了灵活的室外区域。

大楼的临街立面设计是为了响应周边中世纪建筑的立面形式,及其变化的方向。两种不同方向的建筑在此相交,以雕刻手法凸显了独特的街区面貌。纵横交错的混凝土柱梁在空间上被扭曲,强调了建筑立面的厚度。

### 零浪费

这个大楼用轻巧的清水混凝土板建成。这种材料随处可见,使整个大楼看起来里外衔接非常自然流畅。

50cm厚的实心墙全部使用轻巧的混凝土板,避免额外使用保温层或保温膜。这种垒实心墙的办法在中世纪就开始使用,主要用来抵挡恶劣的环境,实现保温效果,主要做法是把木料、稻草和黏土混在一起,制成网状的混合物。现如今,轻巧的混凝土板发挥着相似的作用,只是更方便拆除。材料可完全回收,达到了零浪费的标准。

大楼沿着30m深的界墙伸展,其前后部分通过狭长的中间区域连接。每间公寓内部都是两两相对的布局。楼梯是大楼唯一的共用部分。开放的楼梯和楼梯平台使居民得以时常碰面。富有表现力的外表面在公共和私人空间、三维的外立面、简洁开放的内部空间之间起到了调节的作用。

### House 27D

The project incorporates the historical formation of the surrounding buildings with its 30m long plots and projects this format into the 10m wide front.

The mixed use building completes the city block towards the high street; the courtyard provides daylight from the south where an arcade offers flexible external areas within the otherwise dense historic city center.

The main elevation responds to the punctuated facades of the neighboring medieval buildings and the directional change of

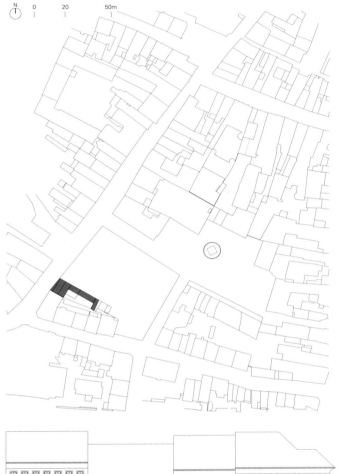

历史立面 historical elevation

 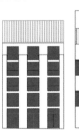 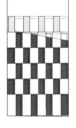 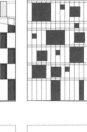 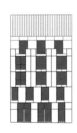  

 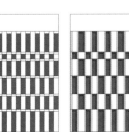 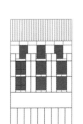 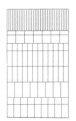 

立面研究 elevation study

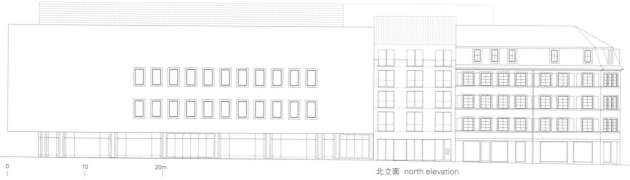

北立面 north elevation

those. An intersection of these two directions emphasizes the perspective in a sculptural manner. A grid of horizontal and vertical concrete columns are spatially distorted thus reinforcing the depth of the facade.

### Zero Waste

The building is constructed of lightweight fair-faced concrete. The material can be seen everywhere thus giving an impression of a fluid transition between outside and inside.

The solid 50cm thick walls are entirely constructed of lightweight concrete, avoiding the use of additional thermal insulation or membranes. They reflect medieval techniques of building solid walls which perform as a weather membrane and thermal mass storage by using a mesh of timber, straw and clay. The lightweight concrete wall performs in a similar way and can be easily demolished. The material is completely recyclable and achieves zero waste requirements.

The building extends along the party wall of the 30m deep site where an elongated area in the middle connects the front and the rear of the building. A vis-à-vis situation is created within each apartment. The staircase forms the only communal area of the building. The open stair and landings invite the residents to meet. The expressive surfaces mediate between public and private areas, the three dimensional external facade and the plain and open internal spaces. Kraus Schönberg Architects

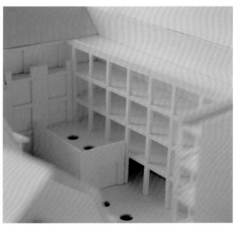

项目名称: House 27D
地点: Constance, Germany
建筑师: Kraus Schönberg Architects
结构工程师: Fischer & Leisering
机械工程师: Greiner Engineering
用途: mixed use building
总建筑面积: 930m²
供热方式: gas, solar thermal
能源使用: 70kwh/m²
造价: EUR 1.5million
竣工时间: 2011
摄影师: ©Ioana Marinescu

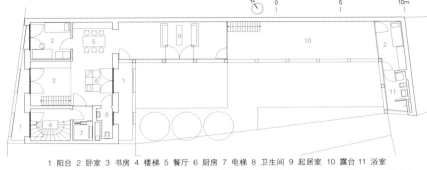

1 阳台 2 卧室 3 书房 4 楼梯 5 餐厅 6 厨房 7 电梯 8 卫生间 9 起居室 10 露台 11 浴室
1. balcony  2. bedroom  3. library  4. stair  5. dining room  6. kitchen  7. lift  8. toilet  9. living room  10. terrace  11. bath room
四层 fourth floor

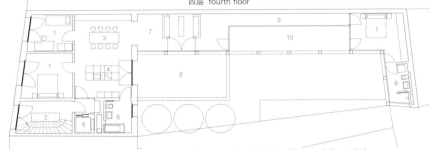

1 卧室 2 楼梯 3 餐厅 4 厨房 5 电梯 6 浴室 7 起居室 8 露台 9 走廊 10 阳台
1. bedroom  2. stair  3. dining room  4. kitchen  5. lift  6. bath room  7. living room  8. terrace  9. corridor  10. balcony
三层 third floor

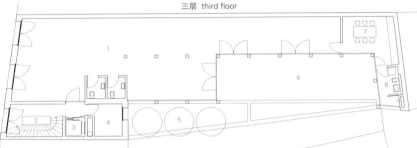

1 零售空间 2 楼梯 3 电梯 4 杂物间 5 花园 6 露台 7 员工房间 8 浴室
1. retail  2. stair  3. lift  4. utility  5. garden  6. terrace  7. staff  8. bath room
二层 second floor

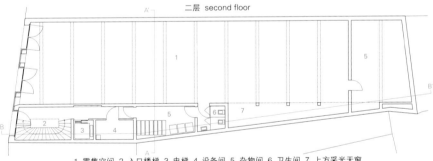

1 零售空间 2 入口楼梯 3 电梯 4 设备间 5 杂物间 6 卫生间 7 上方采光天窗
1. retail  2. entrance stair  3. lift  4. services  5. utility  6. toilet  7. skylight above
一层 first floor

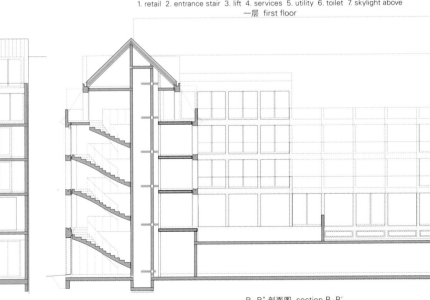

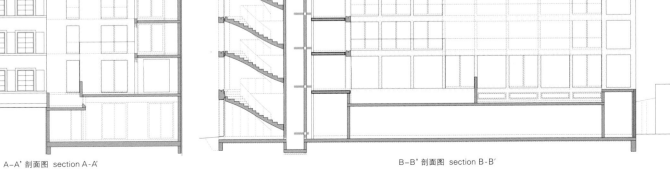

A-A' 剖面图  section A-A'     B-B' 剖面图  section B-B'

# 墨西哥城西班牙文化中心

JSª Arquitectura + Arquitectura 911SC

待完成的西班牙文化中心项目提案要强化和突出文化中心已有的用来承办各种活动和展出的空间功能和作用。文化中心现行的动态模式和文化设计提案必须得到加强,为艺术活动创造空间,拓展方式方法,艺术活动目前还无法在这里举办,但是如果举办就一定要更高效、更吸引人才行。

此项目的主旨之一是构成墨西哥城历史中心城市结构的一部分,使西班牙文化中心发挥桥梁作用,衔接历史中心的不同区域,并在这里发起不同的文化项目。人们可以通过西班牙文化中心从危地马拉大街穿到Donceles大街,同时可以经过展区,让公众可以充分利用文化中心的内部通道。

该项目希望建造一座几层高的大楼,可以因地制宜用来做展厅、会议室、研讨室、工作坊、儿童活动室、电影放映室和音乐厅等。要实现上述目标,最主要的挑战之一是考虑清楚该地段的复杂性。

我们提出的解决方案是在室内不采用中间立柱,屋内空间高于4.6m。

从建筑学角度而言,考虑建筑的历史背景是另外一大挑战。该建筑要融入历史背景,同时要考虑一个是属于现代的,另一个具有历史性,我们提议外立面和建筑体量要与街道和周围的大楼规模互相呼应。混凝土和耐候钢构成方格和凹槽状,仿照市中心大楼的墙面颜色和厚度,同时始终保持建筑的现代化特色。

## Spain's Cultural Center in Mexico

The project to complete Spain's Cultural Center (CCE) proposes enhancing and consolidating the spaces the CCE is already using for a diversity of events and exhibits. The current dynamics and cultural proposal of the CCE must be magnified, creating spaces, ways, forms or formats for artistic activities that at this time cannot be held here, where these activities can be carried out more efficiently and be more attractive.

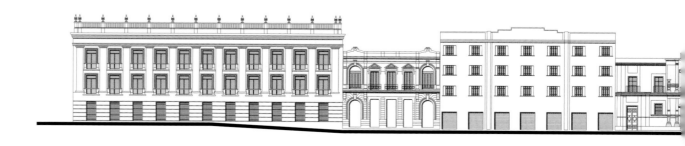

北立面 north elevation

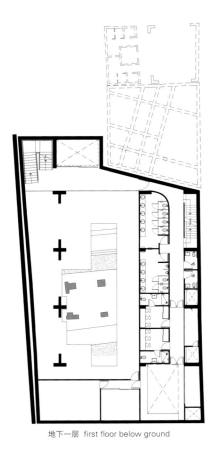

地下一层 first floor below ground

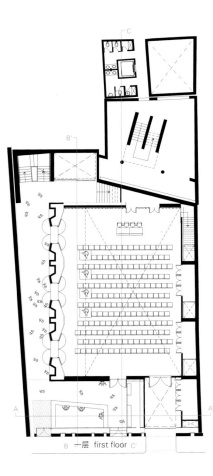

一层 first floor

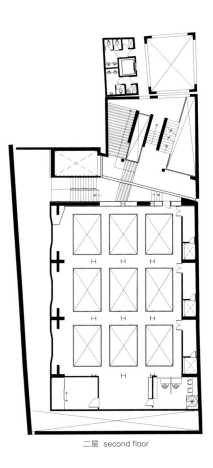

二层 second floor

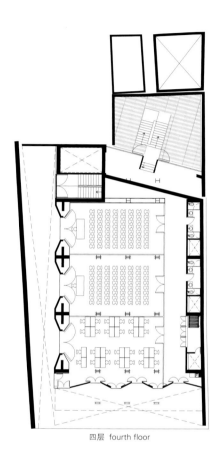

四层 fourth floor

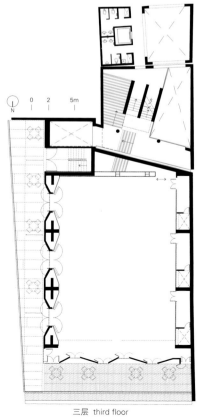

三层 third floor

One of the central ideas of the project is to form part of the urban fabric of Mexico City's Historic Center, considering the CCE as a passage linking different zones and different cultural programs sponsored by the Historic Center. A person could cross from Guatemala street to Donceles street through the CCE, and make this crossing a walk through the exhibit sites, giving this interior passage a strong public dimension.

The program envisages several floors with flexible uses that can work as exhibit areas, conference rooms, for seminars, workshops, children's activities, theater, concerts and so forth. Reaching this objective, taking into account the complexity of the location for the project, is one of the most significant challenges.

The solution we propose is a structure that will permit having large areas without intermediate columns and with heights that go beyond 15 feet.

From the architectural point of view, the relationship with the historic context has represented another major challenge. To integrate the building to this context, one of which is contemporary and the other historic, we have proposed that the facades and the volumes respect the proportions of the streets and buildings that define them. Apparent concrete and corten steel, with lattices and recesses, seek to replicate the colors and depth of the facades of the downtown buildings, always remaining faithful to the contemporary side of the project. JSª Arquitectura + Arquitectura 911SC

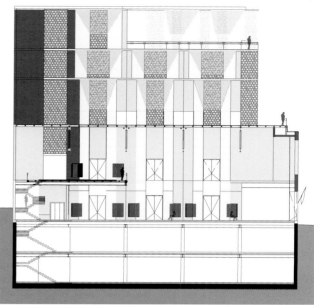

A-A' 剖面图 section A-A'

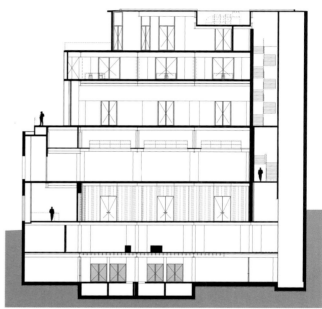

B-B' 剖面图 section B-B'

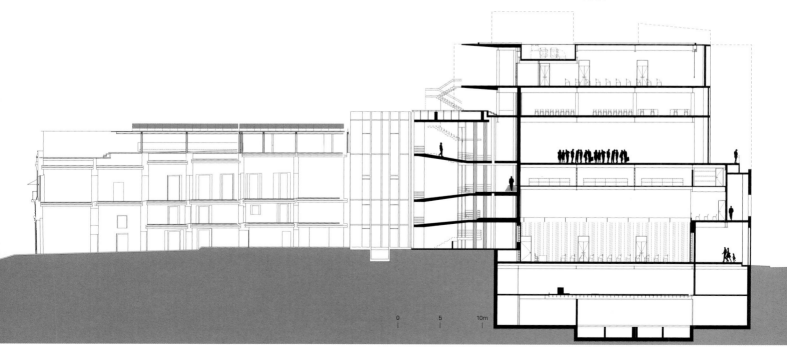

C-C' 剖面图 section C-C'

项目名称：Spain's Cultural Center in Mexico
地点：Guatemala 18 and Donceles 77, Mexico City
建筑师：Javier Sánchez in collaboration with Arquitectura 911 SC
项目团队：Jose Castillo, Saidee Springall, Juan Manuel Soler, Juvencio Nuñez, Pablo Zamudio, Edgar González, Gabriela Delgado, Gustavo Rojas, Domingo Granados, Mariana Paz, Jimena Antillón
甲方：Spain's Cultural Center in Mexico, Embassy of Spain in Mexico
总面积：4,000m²
竣工时间：2010
摄影师：©Jaime Navarro(courtesy of the architect ) - p.113 bottom
©Rafael Gamo(courtesy of the architect ) - p.110, p.111, p.112, p.113 top, 115 top, bottom, p.116, p.117 top
©Moritz Bernoully(courtesy of the architect ) - p.114, p.115 middle, p.117 bottom

# 梦想商业中心酒店
Handel Architects

梦想商业中心酒店是一家精品酒店，位于纽约市切尔西街区，占地17 000m²。共有12层，共计316间客房、两间餐厅、屋顶休息室和VIP休息室、室外游泳池和酒吧、健身房、活动场所和底层零售空间。

梦想酒店选址艰难，店面面向第16和17号大街，西面毗邻海事酒店。1964年，美国国家海事联合会委任来自新奥尔良的建筑师Albert Ledner设计新总部，选址位于第12和13号大街之间的第七大道上。两年后，他又在今天梦想酒店的选址上为总部建了一座附属建筑物。几年后，Ledner先生又在这座附属建筑物上加盖了翼楼，而这个新的翼楼最终被改建成了海事酒店。20世纪70年代，海事联合会解体，大楼被转手用于各种用途。2006年，Handel Architects建筑师事务所决定参与主要附楼的改造工程，打造梦想商业中心酒店。

Ledner先生在1966年为海事联合会设计的附楼富有"差异性"，对其特色的保存也至关重要。沿着第17号大街，倾斜的大楼正面覆盖着不锈钢瓷片，采用了顺砖砌合的方式，和Ledner当年设计原联合会大楼时采用的马赛克瓷砖感觉相仿。新楼立面的窗户如同舷窗的造型，一种与原先大小相同，另一种为原来的一半大小，使原来显得过于僵硬的网格设计看起来比较随意，整个正面显得既排列有序又富有活力。瓷片能倒映出蔚蓝的天空、耀眼的日光及皎洁的月光，当光线恰到好处地照到建筑立面时，不锈钢仿佛不复存在，圆形的窗户造型看上去就像漂浮的泡泡。直角面板在拐弯处折叠，继续保持斜度，与北立面的窗户造型形成了鲜明的对比。

建筑临16号大街的立面当年作为附属楼时是光秃秃的，没有任何装饰，现在焕然一新。表面由两层穿孔不锈钢板铺就，第一层的圆洞和17号大街的舷窗造型如出一辙，里层是规则的钻孔造型。客房外层的防雨层中有序排列着舷窗外形的朱丽叶阳台，地面层向上翻翘，形成了酒店入口的雨篷，展示着酒店的入口。

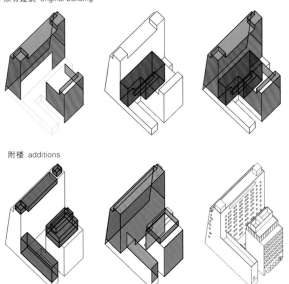

原有建筑 original building

附楼 additions

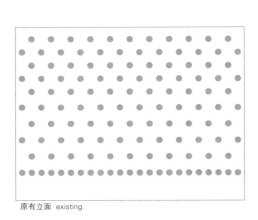

原有立面 existing

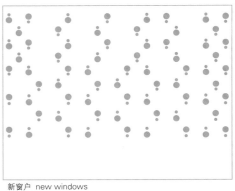

新窗户 new windows

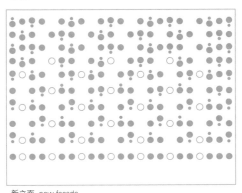

新立面 new facade

原来的设计使整个大楼的自然光效果很差。因此将建筑的中央地带留空,沿着客房的新窗户和阳台打造了新的游泳池平台和躺椅休息处。玻璃底的游泳池使处于大厅的客人可以透过池水看向外面(反之亦然),给人一种飘渺的感觉。采光井镶在大堂、水池和较低楼层之间的柚木框架中,使得整个空间流光溢彩。两百个手工吹制的玻璃圆球吊在大厅里,集中在大理石街餐厅上方,整个空间顿时充满了魔幻气息,吊灯宛若云彩一般。所有的设施和家具都是结合公共空间和客房的需求定制的,和室外的设计相辅相成,持续给客人营造一种空间无限的感觉。

### Dream Downtown Hotel

Dream Downtown Hotel is a 17,000m² boutique hotel in the Chelsea neighborhood of New York City. The 12-story building includes 316 guest rooms, two restaurants, rooftop and VIP lounges, outdoor pool and pool bar, a gym, event space, and ground floor retail.

Dream sits on a though-block site, fronting both 16th and 17th Streets, and is adjacent to the Maritime Hotel, which sits adjacent to the west. In 1964, the National Maritime Union commissioned New Orleans-based architect Albert Ledner to design a new headquarters for the Union, on Seventh Avenue between 12th and 13th Streets. Two years later, he designed an annex for the headquarters on the site where Dream currently sits. A few years later, Mr. Ledner designed a flanking wing for the annex, which would eventually be converted to the Maritime Hotel. In the 1970s, the Union collapsed and the buildings were sold and used for various purposes in the years that followed. In 2006, Handel Architects

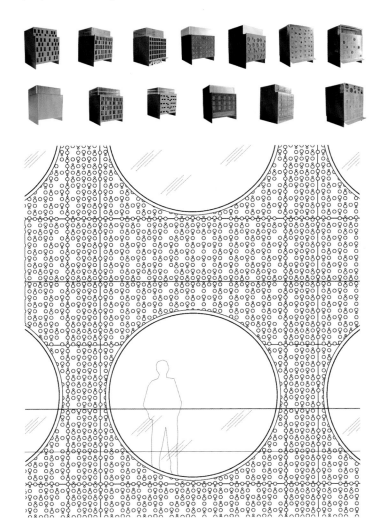

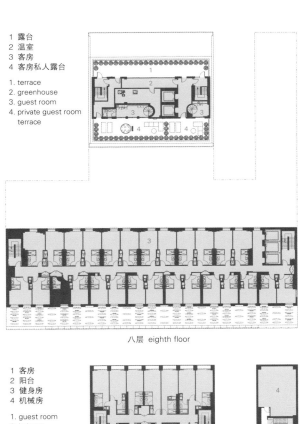

| 1 | 露台 |
| 2 | 温室 |
| 3 | 客房 |
| 4 | 客房私人露台 |

1. terrace
2. greenhouse
3. guest room
4. private guest room terrace

八层 eighth floor

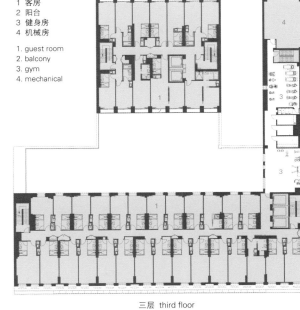

| 1 | 客房 |
| 2 | 阳台 |
| 3 | 健身房 |
| 4 | 机械房 |

1. guest room
2. balcony
3. gym
4. mechanical

三层 third floor

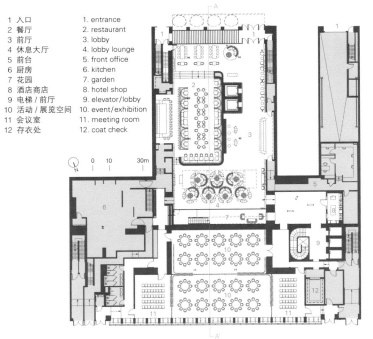

| 1 | 入口 | 1. entrance |
| 2 | 餐厅 | 2. restaurant |
| 3 | 前厅 | 3. lobby |
| 4 | 休息大厅 | 4. lobby lounge |
| 5 | 前台 | 5. front office |
| 6 | 厨房 | 6. kitchen |
| 7 | 花园 | 7. garden |
| 8 | 酒店商店 | 8. hotel shop |
| 9 | 电梯/前厅 | 9. elevator/lobby |
| 10 | 活动/展览空间 | 10. event/exhibition |
| 11 | 会议室 | 11. meeting room |
| 12 | 存衣处 | 12. coat check |

一层 first floor

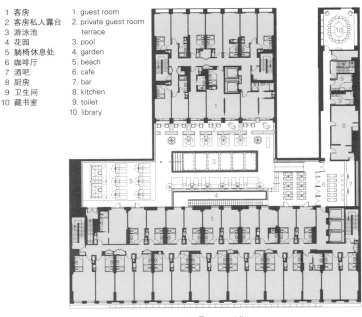

| 1 | 客房 | 1. guest room |
| 2 | 客房私人露台 | 2. private guest room terrace |
| 3 | 游泳池 | 3. pool |
| 4 | 花园 | 4. garden |
| 5 | 躺椅休息处 | 5. beach |
| 6 | 咖啡厅 | 6. cafe |
| 7 | 酒吧 | 7. bar |
| 8 | 厨房 | 8. kitchen |
| 9 | 卫生间 | 9. toilet |
| 10 | 藏书室 | 10. library |

二层 second floor

was engaged to convert the main annex into the Dream Downtown Hotel.

The "otherness" of Ledner's 1966 design for the National Maritime Annex was critical to preserve. Along the 17th Street exposure, the sloped facade was clad in stainless steel tiles, which were placed in a running bond pattern like the original mosaic tiles of Ledner's Union building. New porthole windows were added, one of the same dimension as the original and one half the size, loosening the rigid grid of the previous design, while creating a new facade of controlled chaos and verve. The tiles reflect the sky, sun, and moon, and when the light hits the facade perfectly, the stainless steel disintegrates and the circular windows appear to float like bubbles. The orthogonal panels fold at the corners, continuing the slope and generating a contrasting effect to the window pattern of the north facade.

The 16th Street side of the building, previously a blank facade when the building served as an annex, was given new life. The skin is constructed of two perforated stainless steel layers, its top sheet of holes a replication of the 17th Street punched-window design and the inner sheet a regular perforation pattern. The outer rain screen is punctured with porthole-shaped Juliet balconies for the guest rooms and peels up at the ground level to form the hotel canopy and reveal the hotel entrance.

The original through block building offered limited possibilities for natural light. Four floors were removed from the center of the building, which created a new pool terrace and beach along with new windows and balconies for guest rooms. The glass bottom pool allows guests in the lobby glimpses through the water to the outside (and vice versa) connecting the spaces in an ethereal way. Light wells framed in teak between the lobby, pool and lower level levels allow the space to flow. Two hundred hand blown glass globes float through the lobby and congregate over The Marble Lane restaurant filling the space with a magical light cloud. Fixtures and furnishings were custom designed for the public spaces and guest rooms to complement the exterior design and to continue the limitless feeling of space throughout the guest experience. Handel Architects

项目名称：Dream Downtown Hotel
地点：355 West 16th Street, New York, USA
建筑师：Frank Fusaro
项目建筑师：Elga Killinger
项目团队：Hung Yi Wang, Alan Noah Navarro, Ade Herkarisma, Jacqueline Ho, Luke Lu, Yunhee Jeong, Harshad Pillai, Rick Kearns, John Banks, Jim Rhee, Alexandra Cuber, Danielle Chao, Noelia Ibanez, Slyvie Blondeau, Sang Mi Ji, Barack Pliskin, Vivek Ghimire, Rachel Salatel, Horaci Sanchez
结构工程师：Robert Silman Associates
机电工程师：Thomas Polise Consulting Engineer
土木工程师：Langan Engineering & Environmental Services
外墙：Front Inc
安全与防噪：Shen Milsom Wilke
甲方：Hampshire Hotels & Resorts+Vikram Chatwal Hotels
总建筑面积：17,372m²
竣工时间：2011.6
摄影师：©Bruce Damonte(courtesy of the architect)

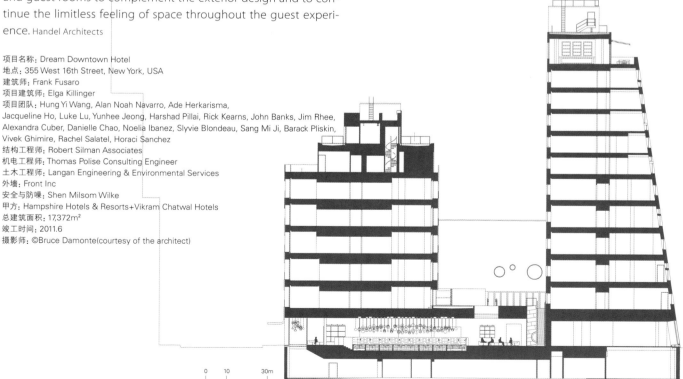

A-A' 剖面图 section A-A'

# 巴黎篮子学生公寓
OFIS Arhitekti

建筑位于巴黎第19区的la vilette公园边缘一块狭长地块，属于由Reichen & Robert建筑师事务所完成的巴黎城市开发项目的一部分。在东北面，新的巴黎电车线路将通过此地，在西南面毗邻电车车库，再边上是一个足球场。建筑的底部三层与电车车库共用一面墙。

学生公寓楼拥有独特的结构：11m宽，南北中轴线上延伸约200m左右，这就意味着正确处理俯瞰着Des Petits Ponts街区延伸地带的东面幕墙十分重要。此外，还将在Des Petits Ponts街修建电车轨道、自行车道和人行道。

这一体量狭长的建筑被分为两个部分，由一座廊桥相连，两部分之间是一座花园。建筑共11层：地下室作为设备用房，一层是公共空间，上面九层都作为学生公寓，布局非常理性和模块化。

项目的主要目的是为学生提供健康的研究、学习和交流的环境。沿足球场边是一个开放式走廊，同时可作为画廊使用，可以俯瞰足球场或是远眺城市景观和埃菲尔铁塔。这个画廊通向学生公寓，为学生提供了一个公共空间。为了优化设计和施工，所有工作室都大小相同，由同样的元素构成：入口、浴室、衣柜、小厨房、工作台和一张床，并且每个小公寓都有一座可以俯瞰整个街景的阳台。

10层楼高的狭长建筑很有存在感。根据功能需要，每个体块都含有两个不同的立面：

面向Des Petits Ponts街区的正立面包含了不同尺度的室外阳台，用HPL木条做外皮。阳台体块随机叠放，使外立面看上去丰富多样且具有韵律感。变化的篮子外观创造了具有活力的外立面，同时消解了建筑的规模和比例给人的体量感。

面向球场的立面包含一条开放的走廊，由3D金属网围挡，工作室入口设于走廊上。两个建筑体块在一层由廊桥连接，廊桥同时也是一个开放的公共空间。

建筑设计高效节能，体现巴黎市为可持续发展所做出的努力。巴黎"气候改善计划"的目标是未来建筑能耗最低标准达到50KWh/m$^2$。为了达成节能和建筑工期两方面的目标要求，此建筑设计力求简洁，全年都达到最佳隔热、通风效果。学生公寓交叉通风，可吸收充足的自然光线。外部走廊和玻璃楼梯设计也提升了自然采光度，既节省了能源，又营造了舒适明亮的社交空间。建筑通过20cm厚的保温材料达到室内保温，在走廊地板和阳台处使用冷桥断路阀，以避免产生冷桥。室内通风采用双流机械通风设计，提供清新洁净的空气，使每套公寓都能全年保持最适宜的温度。屋顶覆盖安装了300m$^2$的太阳能光电板来发电。一个水池用于收集雨水，用来浇灌户外绿色空间。

OFIS Arhitekti

**Basket Apartments in Paris**

The project is located on a long and very narrow site, on the edge of La Vilette park in Paris's 19th district, within an urban development done by Reichen & Robert architects. On the northeast, new Paris tram route is passing along the site. The site is bordering with tram garage on the southwest, above which is a football field. The first 3 floors of the housing will inevitably share the wall with the tram garage.

The parcel has a very particular configuration; 11m in width and extending approximately 200m north-south. This foreshadows the importance of processing the eastern facade overlooking the extension of the street Des Petits Ponts which hosts the tram and both cyclist and pedestrian walkways.

The long volume of the building is divided into two parts connected with a narrow bridge. Between two volumes there is a garden. The building has 11 floors: a technical space in the basement, shared programs in the ground floor, and student apartments in the upper nine floors. The layout is very rational and modular.

The major objective of the project was to provide students with

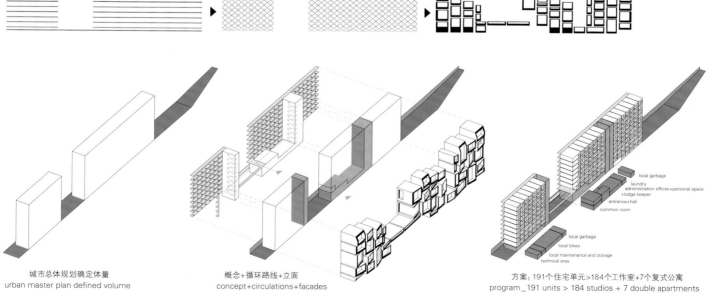

城市总体规划确定体量
urban master plan defined volume

概念+循环路线+立面
concept+circulations+facades

方案：191个住宅单元>184个工作室+7个复式公寓
program _ 191 units > 184 studios + 7 double apartments

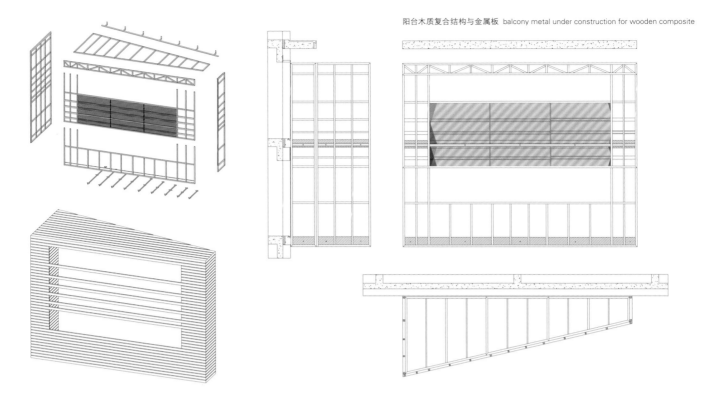

阳台木质复合结构与金属板 balcony metal under construction for wooden composite

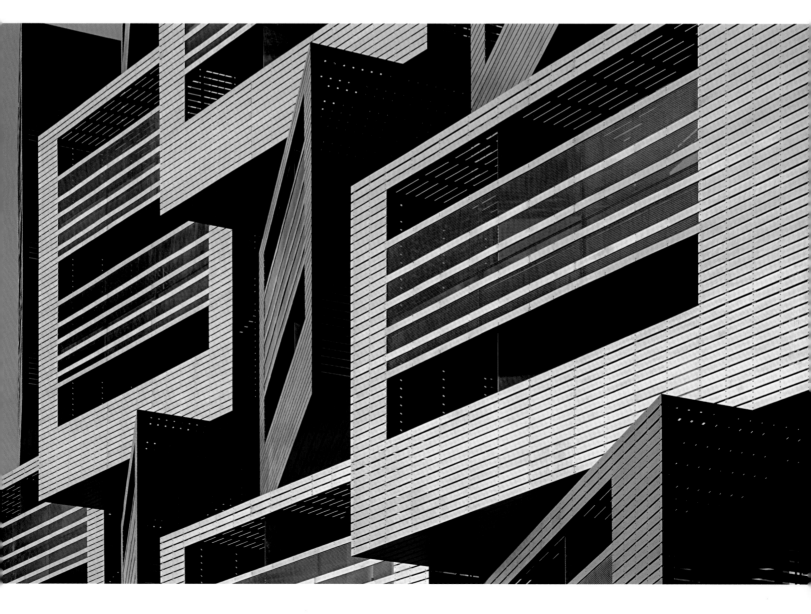

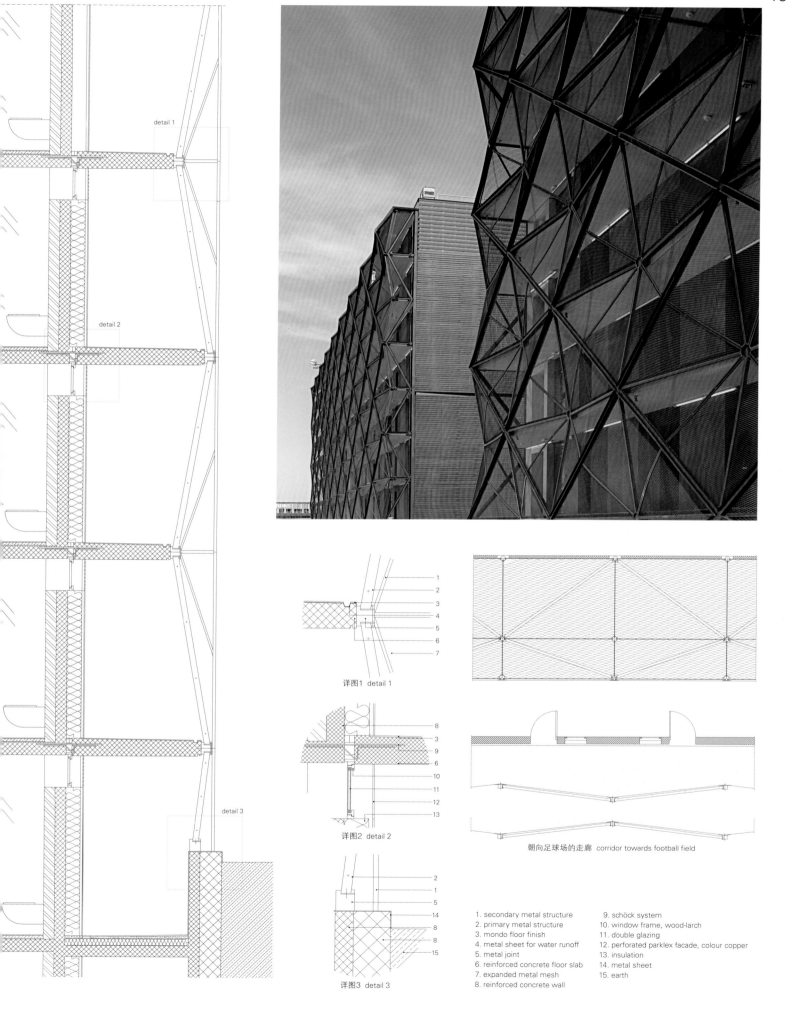

详图1 detail 1

详图2 detail 2

详图3 detail 3

朝向足球场的走廊 corridor towards football field

1. secondary metal structure
2. primary metal structure
3. mondo floor finish
4. metal sheet for water runoff
5. metal joint
6. reinforced concrete floor slab
7. expanded metal mesh
8. reinforced concrete wall
9. schöck system
10. window frame, wood-larch
11. double glazing
12. perforated parklex facade, colour copper
13. insulation
14. metal sheet
15. earth

a healthy environment for studying, learning and meeting. Along the length of the football field is an open corridor and gallery that overlooks the field and creates a view to the city and the Eiffel tower. This gallery is an access to the apartments providing students with a common place. All the studios are the same size and contain the same elements to optimize design and construction: an entrance, bathroom, wardrobe, kitchenette, working space and a bed. Each apartment has a balcony overlooking the street.

Narrow length of the plot with 10 floors gives to site a significant presence. Each volume contains two different faces according to the function and program:

The elevation towards the street des Petits Ponts contains studio balconies-baskets of different sizes made from HPL timber stripes. They are randomly oriented to diversify the views and rhythm of the facade. Shifted baskets create a dynamic surface while also breaking down the scale and proportion of the building.

The elevation towards the football field has an open passage walkway with studio entrances enclosed with a 3D metal mesh. Both volumes are connected on the first floor with a narrow bridge which is also an open common space for students.

The building is energy efficient to accommodate the desires of Paris' sustainable development efforts. The Plan Climates goal is that future housing will consume 50KWh/m² or less. The objectives of energy performance and the construction timetable were met by focusing on a simple, well insulated and ventilated object that functions at its best year round. Accommodations are cross ventilating and allow abundant day lighting throughout the apartment. External corridors and glass staircases also promote natural lighting in the common circulation, affording energy while also creating comfortable and well lit social spaces. The building is insulated from the outside with an insulation thickness of 20 cm. Thermal bridge breakers are used on corridor floors and balconies to avoid thermal bridges. Ventilation is controlled by double flow mechanical ventilation, providing clean air in every apartment with an optimum temperature throughout the year. The incoming air also reuses heat from the exhaust air. The roof is covered with 300m² of photovoltaic panels to generate electricity. Rainwater is harvested on site in a basin pool used for watering outdoor green spaces. OFIS Arhitekti

东立面 east elevation

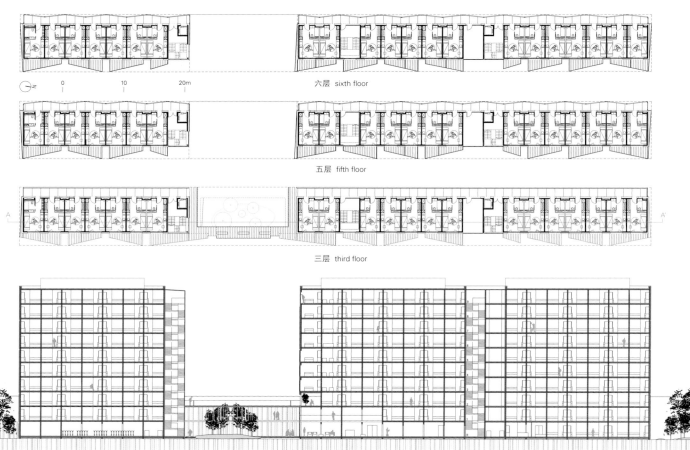

六层 sixth floor

五层 fifth floor

三层 third floor

A-A' 剖面图 section A-A'

项目名称：Basket Apartments in Paris  地点：19th district, Paris, France
建筑师：OFIS Arhitekti
项目团队：Rok Oman, Spela Videcnik, Robert Janez, Janez Martincic, Andrej Gregoric, Janja del Linz, Louis Geiswiller, Hyunggyu Kim, Chaewan Shin, Jaehyun Kim, Erin Durno, Javier Carrera, Giuliana Fimmano, Jolien Maes, Lin Wei
结构工程师：INTEGRALE 4; Bruno PERSON / 机电工程师：Cabinet MTC; Cyril GANVERT / 甲方：Regie Immobiliere de la Ville de Paris
用途：student housing / 方案：apartment studios, common spaces, dining area, living space, storage
用地面积：1,981m² / 建筑面积：studio_35m² building_931m² landscape_1,050m² / 总建筑面积：8,500m²
材料：concrete, glass, metal, plaster, high-density stratified timber panels, expanded metal mesh / 竣工时间：2012
摄影师：©Robert Janez(courtesy of the architect)-p.126~127, p.131, p.132, p.134/©Tomaz Gregoric-p.128~129, p.130, p.135

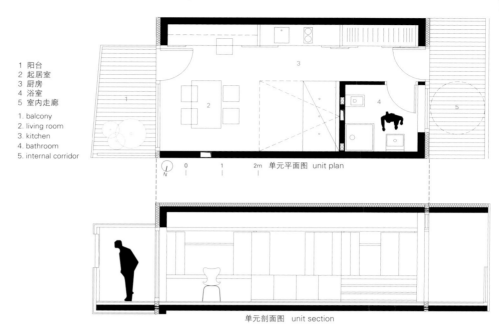

1 阳台
2 起居室
3 厨房
4 浴室
5 室内走廊

1. balcony
2. living room
3. kitchen
4. bathroom
5. internal corridor

单元平面图  unit plan

单元剖面图  unit section

# 层次感活跃了城市表面

# Layering Active

## 重新思考地面层

近期建成的四个公共空间的建筑范例说明了城市景观是如何通过在设计过程中模糊建筑本身与周围环境的界限而增强其社会潜力和外在视觉效果,而成为地标的。

事实上,每座建筑物嵌入地下的部分都增强了它的关系价值,大大提高了其与周围环境的关联性。单纯的内部—外部描述已经不足以形容建筑物与周围环境的多样关系,因此,每座建筑物的地下部分(这一部分也属于内部,与外界环境没有实体上的联系,但仍然与其密切相关)及顶部包围着建筑物的外表面都必须考虑在内。

建筑物的地下部分自成一家,它与外界环境的联系方式常常决定着该建筑物的意向及其在城市景观中扮演的角色。数百年来,公共空间已经成为了其所在社会的代表,同样,如今这些部分埋于地下的公共建筑似乎能最好地阐释这种复杂性和现实的完整清晰度的缺失。

另一方面,一座建筑物的"屋顶"常常会变成另一个"地面",就像一块空白的画布,可以变成公园、广场或者入口。

## Re-thinking the Ground Level(s)

Four recent examples of public space suggest how the urban landscape increases its social potential and visual appeal when its design process blurs the boundaries between the architectural object and its context, becoming a topographical matter.

Indeed, the partial insertion of a building into the ground enhances the relational value of the architecture, multiplying the building's relationships with its surroundings. The dichotomy inside-outside is not sufficient to describe the multiple connections with the context, therefore the buried part (which is an inside with no physical links with the outside, but still in sequence with it) and the surface that encloses the building at the top must be considered.

The underground extension can be a world in itself and the ways it relates with the exterior often define the aspirations and the role of the facility in the urban fabric. For centuries the public space has been the representation of its society, in the same way, nowadays public buildings that are partly buried into the ground seem to describe best the complexity and the lack of full intelligibility of reality.

On the other hand the "roof" of the facility usually becomes another "floor", a blank canvas for possibilities that can be articulated into a park, or a piazza, or a system of access.

---

塞万提斯大剧院/Ensamble Studio
Teruel-Zilla文化休闲中心/Mi5 Arquitectos + PKMN Arquitectura
Georges-Emile-Lapalme文化中心/Menkès Shooner Dagenais LeTourneux Architectes + Provencher Roy+Associés Architectes
南阿尔伯塔理工学院停车场/Bing Thom Architects
重新思考地面层/Simone Corda

Cervantes Theater/Ensamble Studio
Teruel-Zilla/Mi5 Arquitectos + PKMN Arquitectura
Georges-Emile-Lapalme Cultural Center/Menkès Shooner Dagenais LeTourneux Architectes + Provencher Roy+Associés Architectes
SAIT Polytechnic Parkade/Bing Thom Architects
Re-thinking the Ground Level(s)/Simone Corda

# Urban Surface

想要遵照地形特点，创造出一个与原地平面相延续的新的人工地面层是一门技术，该技术属于景观建筑，它使建筑师能够将空间连接起来，并且区分这些空间，使它们具有各自的意义。就一座建筑而言，"地面层"往往能创造出建筑物本身与周围环境之间的功能关系，而其布局大体上能体现整个场地是如何运转的。将建筑物部分埋入地下，而再创造出一个地面层用于修建屋顶，这一做法大大提高了场地多样化发展的可能性，使它可以容纳更广泛的项目，创造城市连接体，此外，这一做法还能增加建筑设计的复杂性和迷人之处。

理解上述理念的一个最佳案例就是最近西班牙特鲁埃尔市修建的Teruel-Zilla文化休闲中心。在这一项目中，Mi5与PKMN两家建筑师事务所合作，完成了对老城广场的改造，以活力四射的全新公众休闲中心取代了陈旧的市场，使周围社区焕然一新。设计者们有意地将建筑物的主体埋入地下以实现两个相互关联的目标：对现存城市景观的尊重和在建筑物屋顶上再造一个公众空间。因此，建筑师将文化休闲中心的屋顶与广场融为一体，成为同一个地平面，在上面设计了一个小型公园和一系列通道。在这个复杂的表面之上，有的部分一直延续到广场，而有的部分则会断开，产生开口、天窗和街道设施，用迷人的外形来吸引人们的注意力。通向地下部分的入口是唯一一处能够揭示出文化休闲中心是建筑而不是景观的部分。与此同时，该入口的吸引人之处在于其优美的线性立面。进入建筑的里面就可以欣赏到它真实的规模——上下共分三层。在这里，红色是主色调，地面、梁及天花板均被涂成了红色，赋予了建筑一种抽象的人工色彩。正因如此，我们可以清晰地看到，建筑的内部设计和外在表现是如何通过使用不同颜色而各具特色的。建筑师们一方面想要通过创造令人愉悦的空间来吸引参观者，另一方面又想通过选用更细微的颜色来使建筑与周围环境融为一体。

墨西哥城的塞万提斯大剧院的入口被作为最有特色的入口而研究，其钢结构的入口与柏林新国家画廊的密斯式建筑风格在一定程度上有些相似。对两座建筑的对比彰显了建筑师的不同意图：德国的设计师意在通过建造讲台和对称布局来强调国家画廊的永恒，而塞万提斯大剧院的设计者Ensamble工作室则构思了一座视觉上轻质的建筑，以经过此地的短暂的人流和太阳为外形。因此，由西班牙建筑师设计的雕塑结构也是扭结的，每个角度所呈现的样子各不相同。在整个项目中，雕塑结构的主要作用是沟通。由于它是一个现代的形式，它所表达出的信息显得迅速而又高度鲜明，而它本身却是一个无言的复杂信息体。事实上，如果我们走进下层——阳光穿过屋顶格栅的斜梁

Acting on the topography with the intention of creating a new artificial ground level in continuity with the existing one is a technique, which belongs to landscape architecture that allows designers to create a connection between places, and to differentiate and give sense to the space. In the case of a building, the "ground level" usually creates the main functional relationships between the building itself and the surroundings, furthermore its layout dictates in the main how the whole site works. Partially burying a building into the ground and layering another ground level to create its roof multiplies the possibilities of the site to host a wider program, create urban links, besides which it increases the potential for a more complex and intriguing architectural design. A recent building that can be taken as example to understand the extent of these themes is the Teruel-Zilla, cultural and leisure center in Teruel, Spain. Here, Mi5 Aarquitectos+PKMN Arquitectura transformed a square of the old city, giving a new life to the neighborhood through the replacement of a dated market hall with a vibrant new public facility. The designers decided to strategically insert the major part of the volume into the ground in order to achieve two goals that are intimately related: the respect of the existing urban fabric and the creation of a public space on the roof of the building. Therefore the architects merged the roof with the square as parts of the same topographic surface, designing on it a small park and as well as a series of passageways. On top of this complex surface, sometimes continuous to the square, other times broken to allow openings, skylights and street furniture polarize the attention of the users with their attractive shapes. The entries to the underground area are the only parts that reveal the cultural and leisure center as a building rather than a landscape. Also in this element the attention for the context is proved by the linear elegance of the facades. Once inside the building, it is possible appreciate the real size of the center, which is articulated into three stories. The red colour is dominant in the material palette; floors, beams, ceiling are painted with it, giving to the facility an abstract artificial look. For this reason it is clear how interior design and external appearance are characterized by a different use of colours. On one hand the architects wanted to engage the visitors through the creation of a playful environment, on the other hand they chose more subtle colours in order to blend the building with its surroundings.

A building whose entry has been studied as the most remarkable feature is the Cervantes Theater in Mexico City. There is a certain assonance between the steel structure that forms its entry and the miesian one of the Neue Nationalgalerie in Berlin. A compari-

位于柏林的新国家画廊，密斯·凡·德·罗设计，1968年
Neue Nationalgalerie, Berlin by Mies van der Rohe, 1968

照射进来，我们就会真切地感受到自已已经置身于剧院之中。参观者的体验就是一个发现的过程，在这个过程中，每一阶段的空间都有和其他阶段的不同之处，因为它们的照明方式、规模及占据其所在地的比例各不相同。矗立在地面上的雕塑架给人一种水平的感觉，这种感觉又通过一个事实得以加强——周围的环境就位于最明亮的部分的下面。参观者进入剧院后的第一感觉是置身于一个上方照明的垂直空间里，然后通过这一狭窄垂直空间中的自动扶梯进入了一个小厅。从外面看，该剧院让人完全难以理解，而实际上，它是整个建筑最后、最精华的部分。

把建筑物建造成一系列不同的空间这一概念自建筑产生伊始就出现了。二十世纪，一些复杂的建筑构成丰富了这一概念，将"走过一座建筑这一行为"转换成了"在头脑中将分离的各部分图像重组或构图"。这些蒙太奇式的转换是半地下建筑常用的方法，参观者对他们进入到内部即将看到的景观几乎一无所知。

这一手法在加拿大蒙特利尔Georges-Emile-Lapalme文化中心的建筑中也有所体现，Menkès Shooner Dagenais LeTourneux建筑师事务所和Provencher Roy及合伙人建筑师事务所共同赋予了这座始建于1972年的文化综合体新的生命。在之前的建筑中，演出大厅、售票处、时装店及餐馆之间的连接空间从未充分实现其潜能。在半地下建筑中，有三个明确的元素：入口、地下空间及其顶部的广场。以上三个主要元素是相互关联的，而且它们看起来仿佛是由共同的表面进行折叠、分割和粘贴而成的不同结构。中心的走廊像折叠的手工折纸一样，标示着进入中心的通道，也是极少的几个标志着广场下面还存在一座建筑的元素之一，其他的元素是用于照明的天窗。内部和外部的连接被压缩到最小，以便使内部成为一个独立的迷人世界，将其抽象化提升到有着高度美感的水平。照明和视听系统、人造天花板以及墙面和地板所用的反射材料，营造出了一个有着高度动感和质感的环境。尽管各个部分之间都有连接的空间来过渡，但界限元素突出了他们之间的变化。将建筑物埋入地下的方式保证了其上面的广场能够尽可能地少受限制——除了偶尔出现的天窗外——从而可以建造一系列

son between the two shows the architects' different approach: the German master wanted to intensify the timeless monumental presence of the gallery, through the creation of a podium and the symmetric composition, while the Ensamble Studio conceived a visually lightweight object, shaped by the ephemeral flows of people passing close by and by the sun. Therefore the sculptural structure designed by the Spanish architects is distorted and presents itself differently from every perspective. Its role in the whole project is mainly communicative and, as in the contemporary form of communications, its message appears quick and highly impressive, however it assumes an unsaid more complex body of information. Indeed, if we follow to the levels below to where the sun's rays seep through the slanted beams of the roof grid we will find ourselves into the theater spaces. The viewer experience is scanned into a discovery sequence in which the spaces of each phase differ from the others because of the way in which they are lit and for dimensions and proportions of the area. Above the ground the sculptural pergola creates a horizontal feeling, amplified by the fact that standing underneath the brightest part are the surroundings. When a visitor enters the facility, firstly he finds himself in a vertical space lit from the top; secondly he accesses the smaller foyers passing through a system of escalators located into a narrow vertical area. The theater, completely unintelligible from the exterior, is placed at the end of this journey as its climax. The construction of the architecture as a sequence of different spaces is a concept that can be found since architecture has existed. During the twentieth century the complexity of some architectural compositions enriched it, transforming the act of walking through a building into a moment of mental re-composition or mapping of separate partial images. These montages are typical of the partially underground facilities, where a viewer have a very little anticipation of what he is going to encounter once inside. This approach can be recognized in the Georges-Emile-Lapalme Cultural Center in Montreal, where the Menkès Shooner Dagenais LeTourneux Architectes and Provencher Roy+Associés Architectes gave a new life to a 1972 cultural complex. In the former building, the connective space among performance halls, box offices, boutiques and restaurants had never fulfilled the potential it had. In the partially underground structure three main elements can be defined: the entry, the subterranean space and the square on top of it. All three elements are continuous and they seemed to be formed by the same surface that is folded, cut and patched with different textures. The porch of the center, which is folded like an origami, depicts the entry and it is one of the few elements that suggest the presence of a proper building under the square; the others are the skylights that bring the light below. The communication between interior and exterior is reduced to the minimal level, hence the interiors can be a separate and fascinating world in which the abstraction is brought to a high aesthetic level. Lighting and audiovisual systems, false ceilings, the use of reflective

有着奇特立面的南阿尔伯塔理工学院停车场
SAIT Polytechnic Parkade with the pixilated facade

1. Ilhyun Kim, "Realta' e Architettura: Totalita' e Dissoluzione dell'Oggetto" in *Spazi Pubblici Contemporanei Architettura a Volume Zero* edited by A. Aymonino and V.P. Mosco, Milano: Skira, 2006

的通道和景观。结果产生了两个独立的连接层：一方面，地下热闹的大厅文化和商业设施与外面的马路相连接；另一方面，顶部的广场仍然维系着建筑物之间的传统联系，满足人们的通行需要，从城市角度来看，这是一个基本方面。

加拿大卡尔加里的南阿尔伯塔理工学院停车场是又一个例子，它展示出一座有着简单的地面造型的建筑能够怎样实现驭繁于简，还给建筑物本身增加功能之外的其他价值。Bing Thom建筑师事务所需要在不破坏宜人景观的前提下为南阿尔伯塔理工学院设计一个35 000m²的停车场，该地景观以一座位于绿色山坡顶端、建于1921年的建筑为特色。该设计方案的关键点在于要将三层的停车场置于现存的足球场之下，使其与老建筑处在同一水平面上并保留通向周围道路的景观长廊。如果是建在地面之上，该建筑的巨大体量一定会非常壮观，但它却与地形融为一体，从而减少了冲击力。通过这种方式，停车场在地面上就像一块岩石，它不但外形合理，而且与地面相融合，看起来就像是景观的自然组成部分。建筑与景观的融合缩小了两者之间的差别，它是通过抽象化的方式来移花接木。这种精神层次的精雕细琢在停车场立面覆盖物上表现得淋漓尽致，从而将停车场变成了一件艺术品。本案建筑师与温哥华的著名艺术家Roderick Quin合作，创造出一个嵌板系统，通过上面的小孔可以看到绚烂的天空。这种人工和自然之间的平衡使停车场具有一种精致的优雅，这在同类建筑中并不常见。为了处理这座建于1921年的建筑，停车场入口的形式、所用材料和规模都与新建筑的其他部分完全不同，就像一个二元的独立构件一样立在地面上。

正如IlHyun Kim所说，当代建筑实践都与未来主义和简约主义艺术密切相关，环境、建筑本身和参观者充分互动，形成一个融洽的"开放作业"。从这方面来讲，将建筑部分埋入地下的项目是对"开放作业"这一概念的最佳解释，由于建筑本身与环境融为一体，参观者想要全面了解该建筑的唯一途径就是在其穿越建筑物的过程中不断地重组自己头脑中记住的画面。

materials for walls and floors, creates a highly dynamic and sensorial environment. Even though the space flows from section to section as a continuum, threshold elements underline the change among them. Keeping the structure underground ensures the square on top is maintained as free as possible, in order to host a series of passageways and landscaped areas, only interrupted now and then by skylights. As a result, two separate levels of connections are established: on one hand in the underground a buzzy foyer links the road with the cultural and commercial facilities, on the other hand the square on top still provides a traditional relation among the buildings and the offers possibilities of passing through, a fundamental aspect from an urban point of view.

The SAIT Polytechnic Parkade in Calgary is another example of how with a simple topographic gesture it is possible not only to address the multiple goals of a brief but also to add value to a facility that could have been merely functional. Bing Thom Architects had to design a 35,000m² car park for the Southern Alberta Institute of Technology without compromising the pleasantness of the landscape, which is characterized by a 1921 building on top of a green hillside. The key point of the site strategy was to localize the three storey parking under the existing football court, enabling it to remain at the same level as the historic building, preserving the view corridor from the surrounding roads. The large facility volume, which could have been imposing if the building was standing out on its own, gets merged with the terrain, reducing its impact. In this way the parking stands out off the ground like a rock and, despite its rational form, the integration with the ground makes it look as if it was a natural component of the landscape from a mass point of view. The process of merging landscape and architecture, reducing the distance between their identities, it is a form of appropriation that passes through abstraction. This mental elaboration is embodied by the skin that covers the facades, which transforms the car park into a piece of art. The architects, in conjunction with Vancouver artist Roderick Quin, created a paneling system that through small holes draws the image of a cloudy sky. The balance between artificial and natural makes the building shift towards a sophisticated elegance, which is uncommon in buildings of this type. To address the presence of the 1921 building, the car park entry has a form, materials and scale completely different from the rest of the new building and projects itself from the ground as a binary isolated element. As noted by IlHyun Kim, the contemporary architectural experience has a close relationship with the futurist and minimalist art, in which environment, architectural object and viewer interact to form a perfect "open work". Following this perspective, projects in which the building is partially buried into the ground are the best expression of the "open work" concept, because context and building are amalgamated and the only way a visitor can have a clear view of them is to recompose the multiple images that he collects in his mind while he is walking through them. Simone Corda

# 塞万提斯大剧院
Ensamble Studio

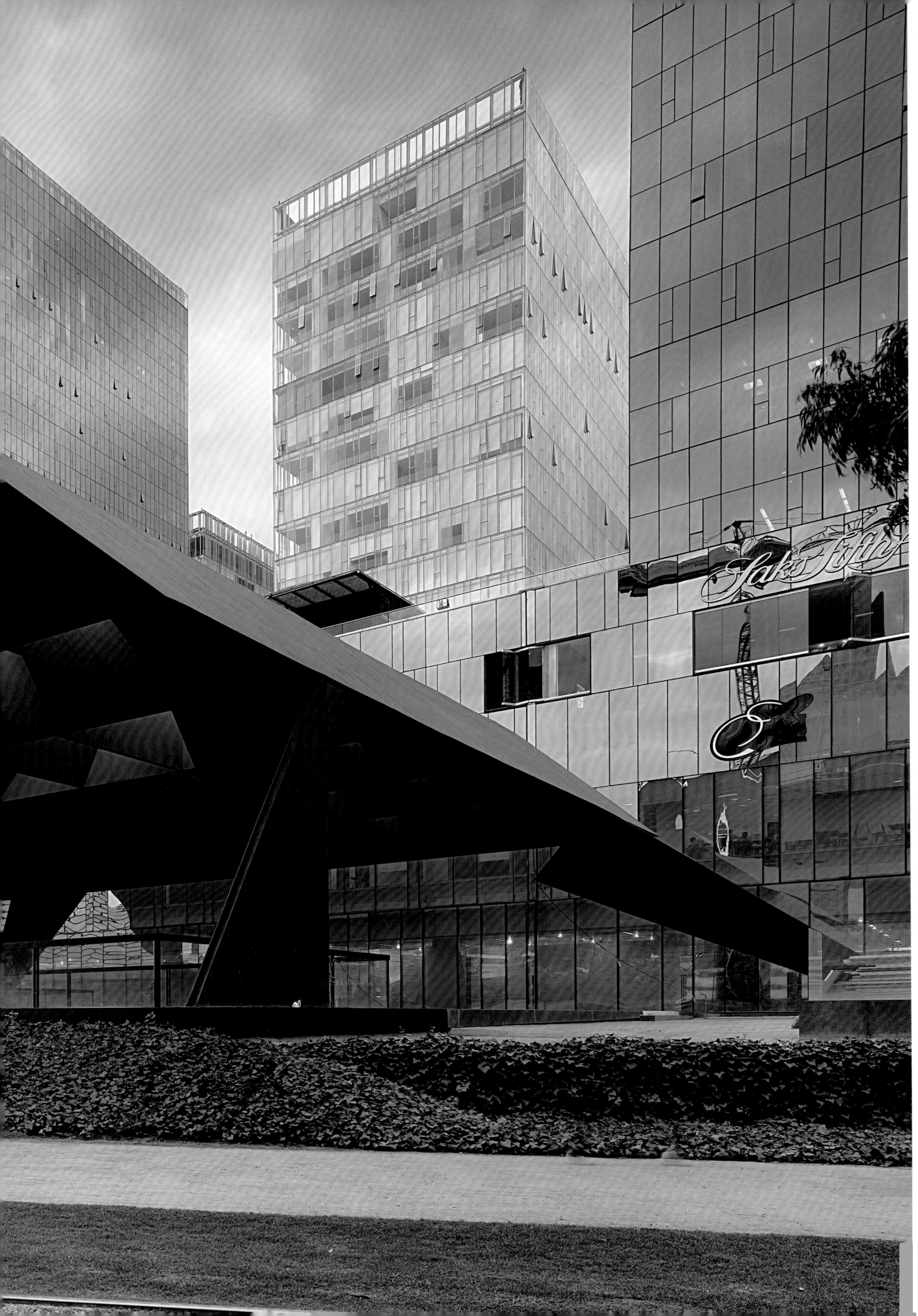
Saks Fifth

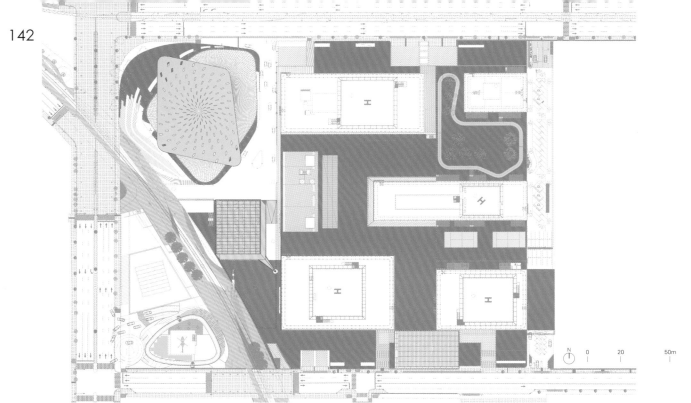

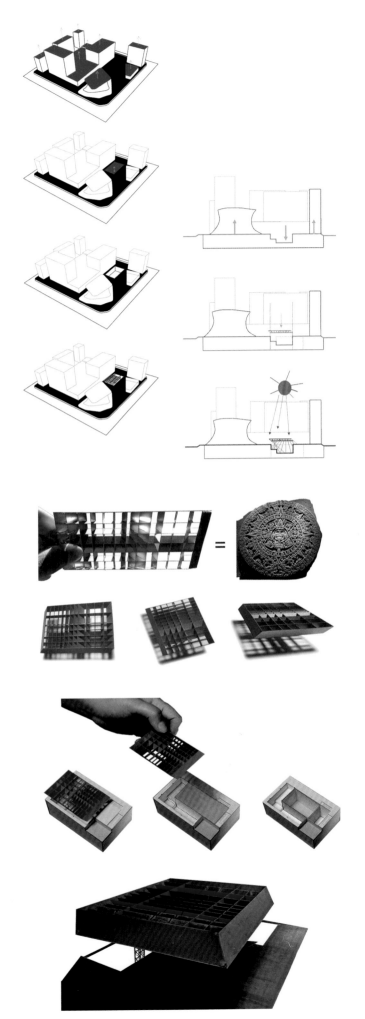

拱顶石由气泡石构成,又称为"太阳石"。巨大的阿兹特克玄武岩开凿于墨西哥城的Zócalo,意为"日之运行",代表着太阳神托纳季乌,他手里抓着心脏(根据阿兹特克神话,托纳季乌成为太阳以后,向诸神索命,诸神于是献出心脏),代表着对日光的永恒需要。我们在这块精美的文物上所看到的光线就是光的象征,这只能通过我们自己发现"我们是什么""我们感觉到了什么"和"我们做什么"来寻找。这就要求我们要有恒心和耐心(土)、精神力量(火)、对不同的生活环境的适应能力(水)和谨慎(气)。难怪不管是阿兹特克人、印加人、玛雅人,还是埃及人等,都将太阳视为生命中唯一的精神支柱,希望将其外在特征与精神特征相联系,从而探索巨大的无形世界。

建筑常常关系到无形世界,然而它又经常使用一些具有震撼效果的具体元素来与无形世界关联。从金字塔的材料中,我们可以看到墨西哥文化中建筑、社会、城市以及象征物的壮丽景象。在与其相邻的玛雅文化中,天然井是一种地貌结构,挖掘出来的空间是神圣的象征,向阳光和雨露敞开胸怀。

根据我们的理解,当代建筑持续地表达着时间推移的延续性。历史的积淀融合并交织了墨西哥文化,也给了当今建筑巨大的灵感。因此,拱顶石由气泡石构成,由一系列开凿的平台空间支撑,当阳光经过其横隔时就会产生时光推移的效果,并且还能为我们遮风挡雨。顶拱石尽可能地收集其上方空间的共振,并使它们有序化。开凿的空间向公众开放,并面向天空,由具有象征意义的钢结构保护着。项目面临着其所用的建筑元素自身的问题:垂直的特点导致了负空间的密度问题,拱顶石所容纳和支撑的空气带来的水平张力问题。拱顶石是一个抽象平衡的关键部位,它减轻了重量,看起来像云朵一样飘逸、多变而轻盈,通过过滤阳光,提升了地面的品质。

然而,空间有自己的规则,建筑就是要巧妙地处理这些规则。建筑结构的设计建立在调和各向同性的内弧面、正交和双向顺序的矛盾的基础上,本质是其组成部分多变的几何形状。将精确的顺序转化为自由的空间而非压制它,使得自然元素(水、光、空气)能够影响到建筑的最终构成。我们看到的现实是表面上的顺序与波动的空间进行博弈的结果,这种博弈使结构板条有差异地暴露于阳光中,形成四个光照强度不同的区域,成为四个开凿的空间的投影,使大厅的螺旋形更清晰。而后又是四时、中美洲关于创造世界的神话中的四个太阳、自地球诞生之初的四个阶段……

它们中任何一个顺序都无法表达太阳与地球之间的联系,因此在设计使拱顶石与地面之间产生重力连接的结构时,本案建筑师决定采用一个非常形象的设计,仿佛它就是女像柱和大力神一样,来唤起多样性,反映出即将住在其中的人们的生活。因此,在设计时也将立柱包含其中,它们依靠刚性结构,随意地分布在空间中,在拱顶石的网格里呈不对称分布。

当我们沉浸在开凿大厅那开放的负空间里时,我们有一种由始而终的感觉。一旦进入地下空间,大剧院仿佛是这个空间序列的尽头。在这里,建筑的融合达到了登峰造极的地步,有着使时间停止、让人们娱乐和沉思的功能。

## Cervantes Theater

The Dovela (Keystone) is an air stone, the "Sun Stone." The Aztec basalt monolith excavated in the Zócalo in Mexico City, means "Tonatiuhtlan de Ollin" or "Sun of Movement." The god of the sun it represents, Tonatiuh, grabs a heart and expresses the need for continuity of solar time. The rays we can appreciate in this beautiful archaeological piece are the symbol of light, which we have to find through the discovery of what we are, what we feel and what we do. Tenacity and patience are required (Earth), also spiritual strength (Fire), capacity to adapt to different circumstances of life (Water) and mindfulness (Air). No wonder the Aztecs, Incas, Ma-

项目名称：Cervantes Theater
地点：Mexico D.F., Mexico City
建筑师：Antón García-Abril
合作建筑师：Elena Pérez, Alba Cortés
合作者：Débora Mesa, Joaquín Gallegos, Alba Beroiz, Jaime Alcayde, Cristina Moya, Juan Ruiz Antón, Tomaso Boano, Federico Letizia
结构工程师：Colinas De Buen
建筑面积：11,500m²
竣工时间：2012
摄影师：©Roland Halbe

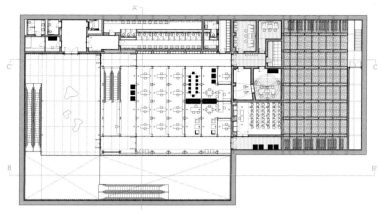
地下一层 first floor below ground

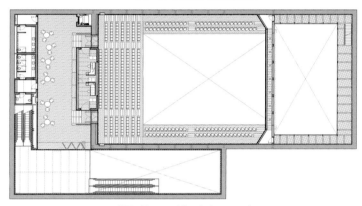
地下二层 second floor below ground

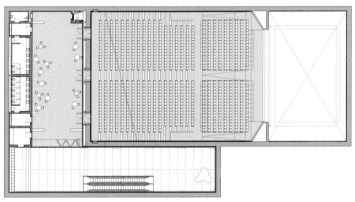
地下三层 third floor below ground

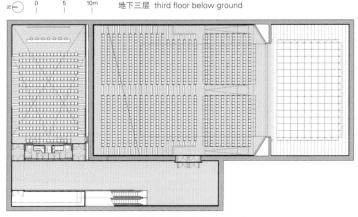
地下四层 fourth floor below ground

yans, Egyptians, etc., identified the sun with the universal spirit of life, trying to associate its physical characteristics with the spiritual ones. Thus, they would reveal the greatness of the intangible. Architecture refers to the intangible constantly. However it does so using elements of imposing physicality. The pyramids, as material eruptions, show a great constructive, social, urban and symbolic splendor in the Mexican culture. And in the neighboring Mayan culture, the cenotes are geomorphologic structures, where the sacred is the excavated space, open to light and rain.

We understand contemporary culture as a constant expression of connectivity with the movements of time, and the layers of history that overlap and hybridize the Mexican culture are of great inspiration to make a work of architecture today. Therefore, the Dovela appears as a stone of air, supported by the space that comes from a sequence of excavated terraces; that offers to the sun which moves the time when going through its slats; and that protects from the rain and shelters us inside the earth. The Dovela tries to collects all the resonances of the world emerging above it, to give them order. The excavated spaces are given to the public and open to the sky, protected by the symbolic metal structure. The project confronts the elemental natures with which it is built: the deep density of the negative space, of vertical character; and the horizontal tension of the air contained and supported by the Dovela, last key piece of an abstract balance that loses its weight to appear aerial, mutable and light as a cloud that qualifies the space in the ground by filtering sunlight rays.

However, space dictates its rules, and architecture manipulates them. The design of the structure is based on the paradox of reconciling the isotropic order of its intrados, orthogonal and bidirectional, with the essence of the variable geometry of its section. A mathematical object which transposes its strict order to the space but does not impose it, allowing the natural elements (water, light and air) to affect its last configuration. The perceptual reality is the result of this struggle between the apparent order and the fluctuating space, that causes the exposure to the diversity of the solar lights that vibrate between the slats of the structure, creating four fields of intensity which are the projection of the four excavated spaces that articulate the spiral of the theater lobbies. Again the four times, the four suns of the Mesoamerican legend about the creation of the world, the four stages of the Earth since its creation…

The contact between the sun and the earth could not pick up patterns of the order of any of them. So in the design of the structure that generates the gravitational connection between the Dovela and the excavations in the ground, we decided on a very figurative design, as if it were caryatids and Atlanteans; evoking diversity, reflecting the world of the man that will inhabit the space. Therefore, the pillars are involved in the movement, in time, and are freely arranged in the space dance, relying on the rigidity of the structure its non symmetric disposition within the internal grid of the Dovela.

By immersing ourselves in the negative space of the open excavated lobbies, we can have access to a new and final happening. Once inside the earth, the Theater appears as the end of this sequence of spaces. Here the synthesis of the building culminates with the function of a halted time, recreated, a place to contemplate. Ensamble Studio

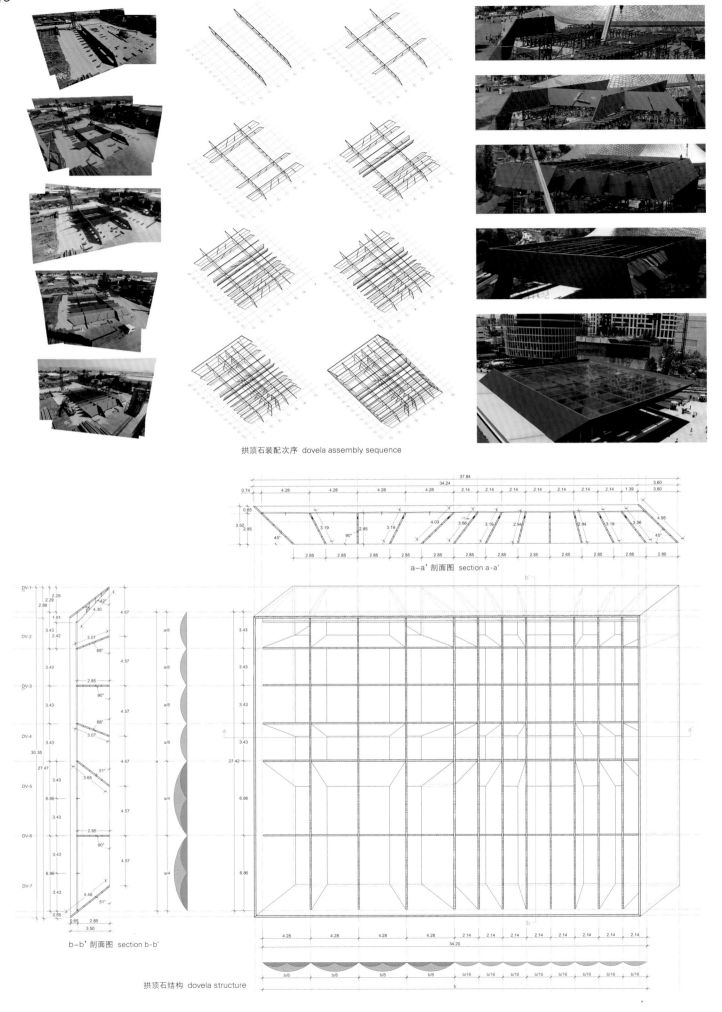

拱顶石装配次序 dovela assembly sequence

a-a' 剖面图 section a-a'

b-b' 剖面图 section b-b'

拱顶石结构 dovela structure

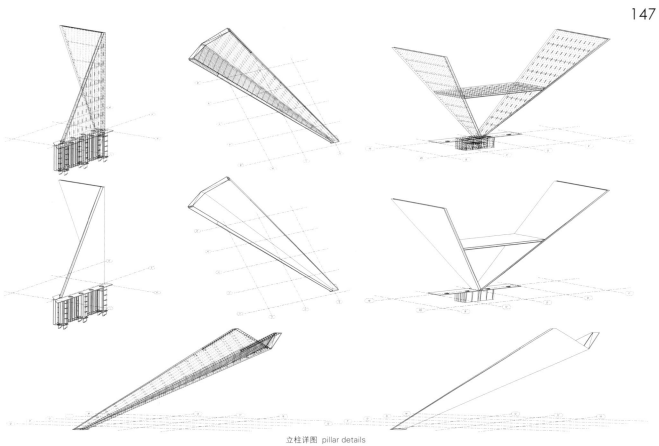

立柱详图 pillar details

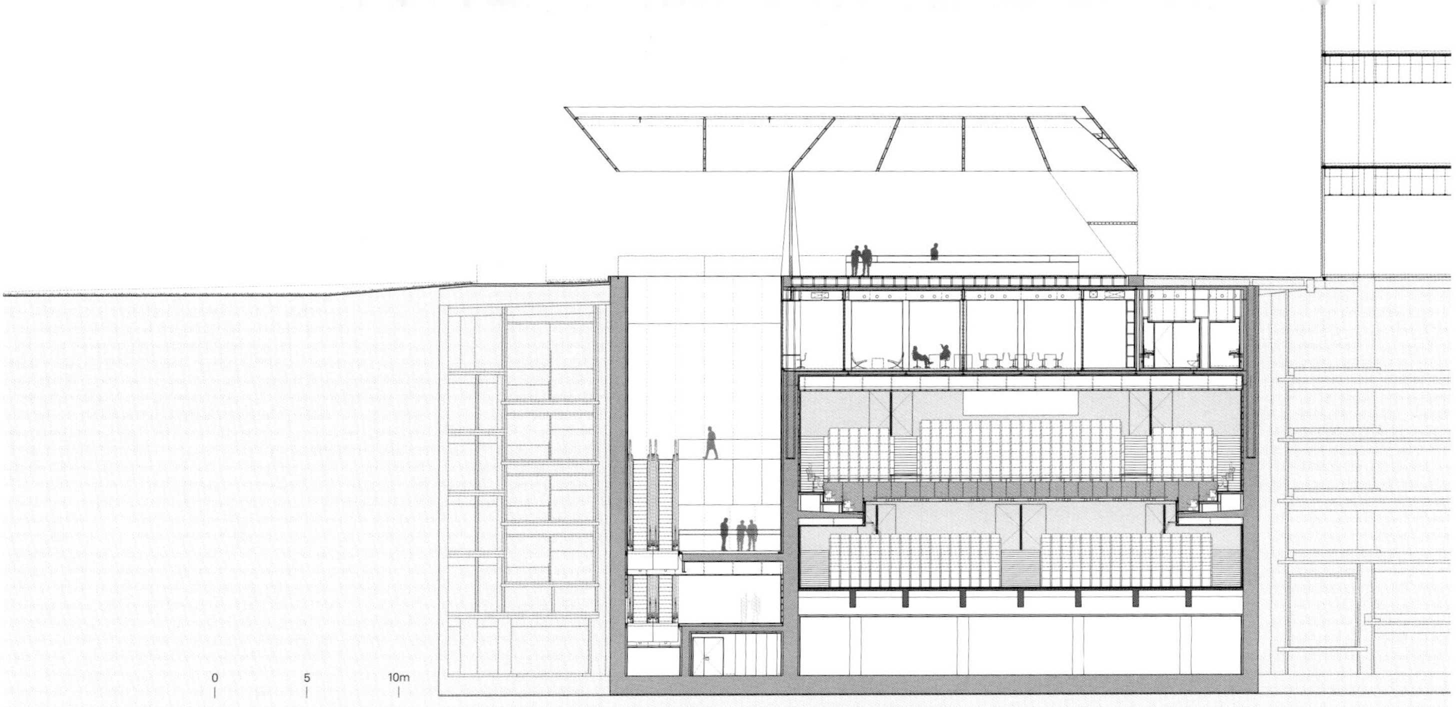

A-A' 剖面图 section A-A'

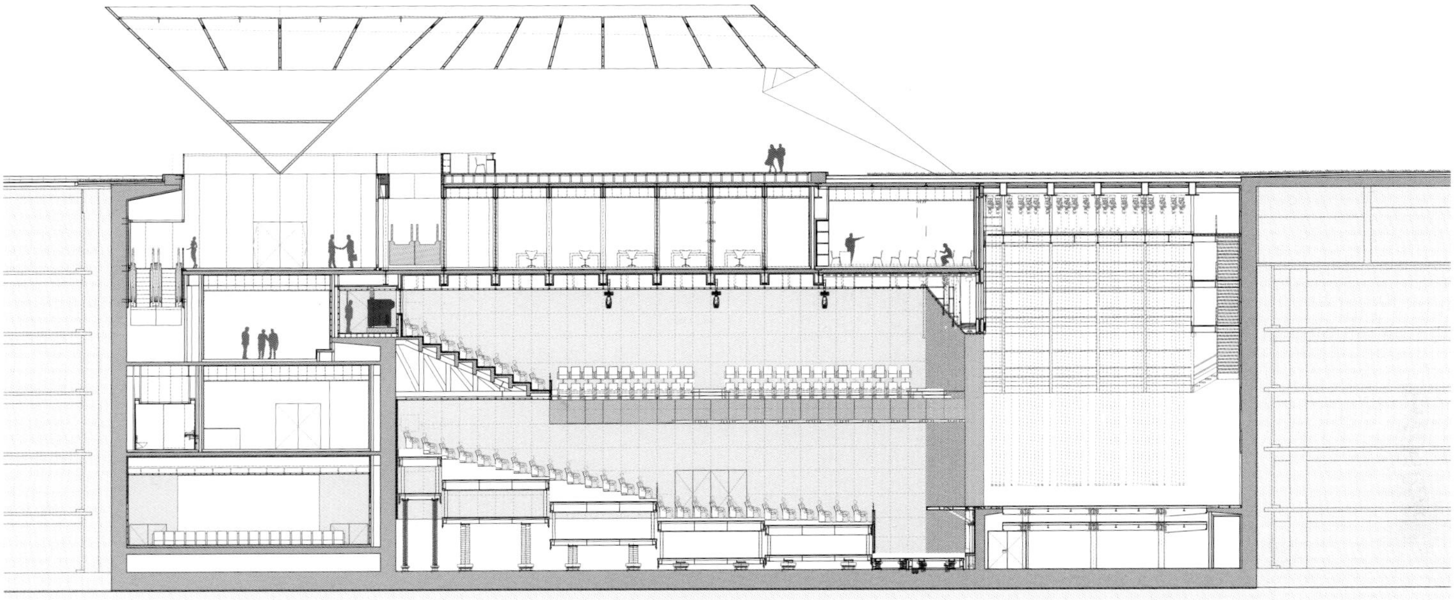

B-B' 剖面图 section B-B'

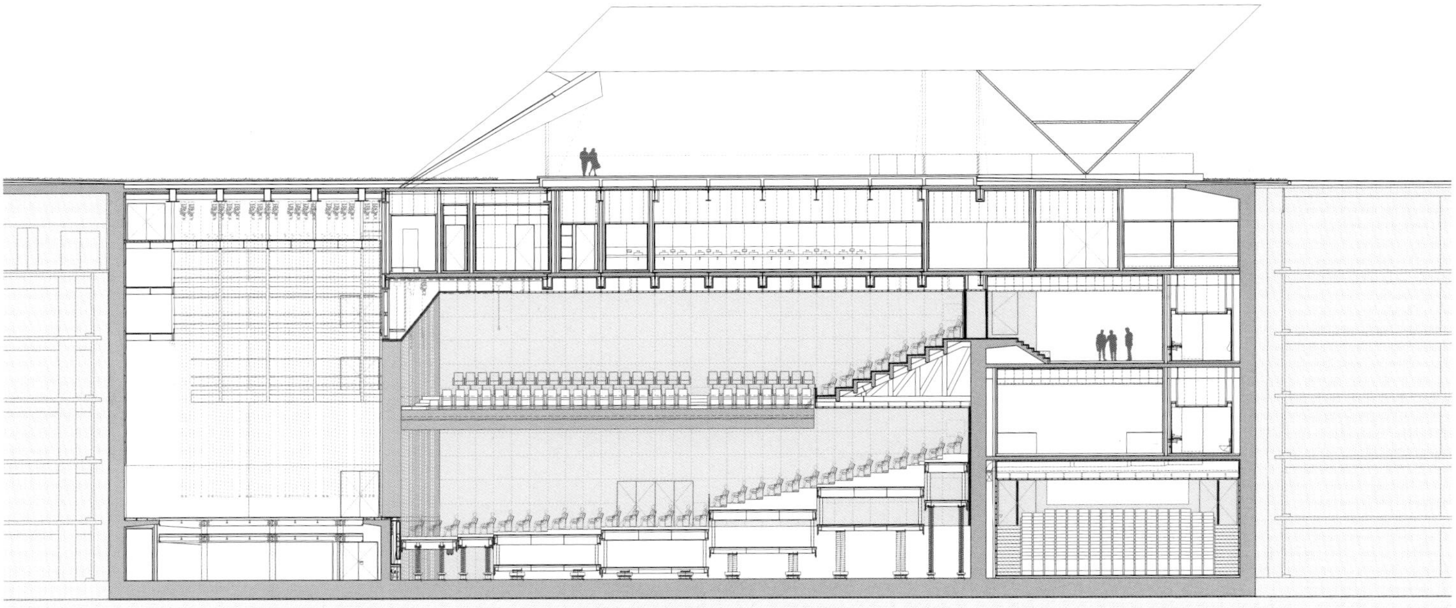

C-C' 剖面图 section C-C'

# Teruel-Zilla 文化休闲中心
Mi5 Arquitectos + PKMN Arquitectura

城市设计 | 层次感活跃了城市表面 | UrbanHow Layering Active Urban Surface

### 地下休闲中心和公共空间

1987年,由Sanz、Buscalioni、Casanovas和Santafe等组成的研究小组在特鲁埃尔省加尔韦镇挖掘出了阿拉贡恐龙遗迹,并进行了编目。这是首次在伊比利亚半岛发现恐龙。

特鲁埃尔的地下充满了令人新奇的发现,为我们重现了特鲁埃尔省产生之初的那些强大的生命,这正是Dinópolis主题公园一直就想传播的。让人惊奇的是,我们必须要返回它的最深处去激发它的活力。

面对陈旧的市场建筑占据大部分Domingo Gascon广场公共空间的现状(尽管还保留着一个纪念雕塑),本案建筑师决定拆除旧的结构(一个呆板而排他的建筑类型),并在地下建造一个体量庞大的青少年休闲活动空间来振兴和促进特鲁埃尔的活动,而在市场建筑建成之前就存在的旧公共广场将被收回。

公共空间和休闲中心项目就像是埋在地下的哥斯拉式工程,呈现出一种当代流行的风格。

埋入地下的巨大体量推动并挤破地表,制造出一个全新的城市地形。参观者将进入这个已经成为公共广场的地表之下,再往下走,在各层之间享受会议、娱乐和运动场所。新的活动和时代特征显示了新的都市生活方式,尤其对于一个历史占据了重要空间的城市而言。在认为有潜力的地方,建筑师发现新的、可行的表达方式并敢于去实践。这种实验手法是在寻找结构方式和技术规范的界限,探索一种全新的建筑原型,这种原型在有可能为增强市民关系、振兴城市历史区域并改善城市的生活提供新的方式。

### Teruel-Zilla

#### Underground Leisure Lair and Public Space

In 1987 the remains of "Aragosaurus ischiaticus" were dug up and catalogued by a group of researchers consisting of Sanz, Buscalioni, Casanovas and Santafe at the town of Galve, in the province of Teruel. It was the first dinosaur to be found at the Iberian Peninsula.

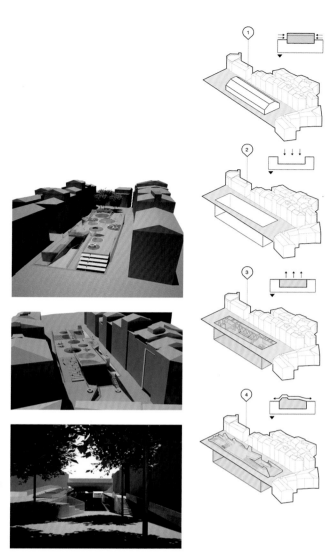

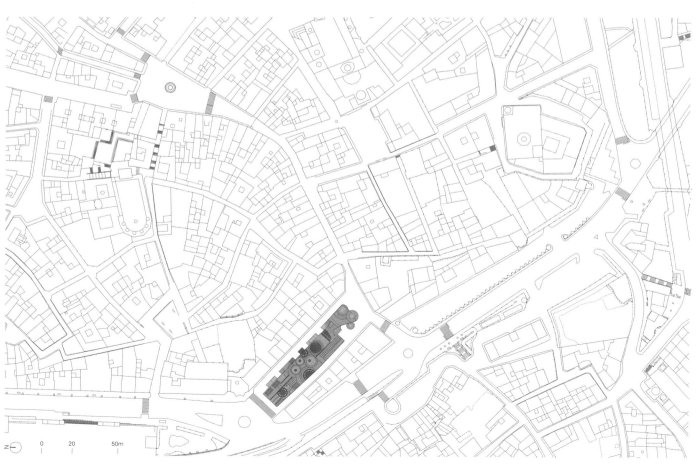

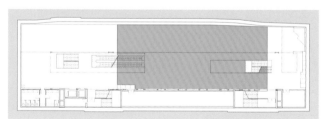

地下一层 first floor below ground

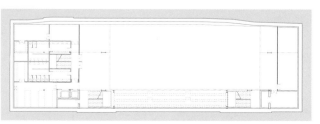

地下二层 second floor below ground

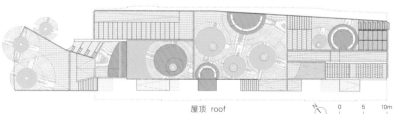

屋顶 roof

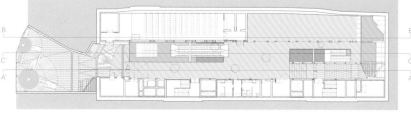

一层 first floor

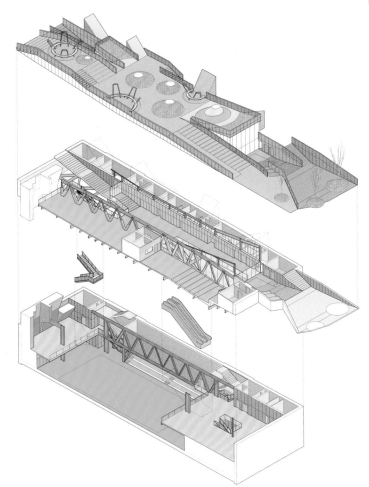

Teruel's underground, as thematic parks as Dinópolis want to spread, is full of discoveries that remind us the lost existence of a powerful life in the origins of the province. It's surprising that we have to return to its deepness to try to reactivate it.

Facing the existence of an obsolete and underused market building occupying most of the meagre public space of the nearly nonexistent Domingo Gascon Square(even if provided with a commemorative statue), the decision taken is that of demolishing this old structure(a very inflexible and exclusive typology) and introducing a huge volume of youth leisure activities on the underground, to revitalize and to foster Teruel's activity, while the old public square existing before the market building was constructed is regained.

The public space and leisure center project takes a buried Godzilla's expression: a telluric element of contemporary and pop expression.

The big buried volume pushes the earth surface till it breaks and produces a new urban topography. The visitants will settle this surface that becomes into a public square, and they will go down in between the stratums, being entertained by meeting activities, fun and sports. The new activities and their contemporary manifestation, make evident new ways of urban dialogue, especially in a city where history has taken up such an important space. It is the discovery of new possibilities of expression and in a daring appropriation of them, is where we consider that its potential lies. This experimental typology is an investigation that pushes the boundaries of structural means and technical regulations to explore architectural prototypes that may be able to generate some new ways in achieving deification, setting an scenario of public enhancement and optimism in order to accomplish citizenship empowering, historical urban fabric revitalization and intensification of city life. Mi5 Arquitectos + PKMN Arquitectura

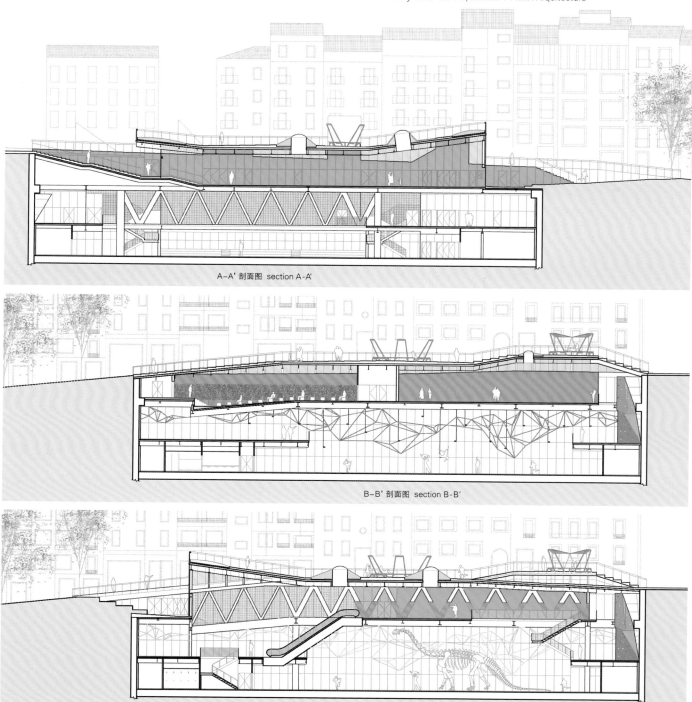

A-A' 剖面图 section A-A'

B-B' 剖面图 section B-B'

C-C' 剖面图 section C-C'

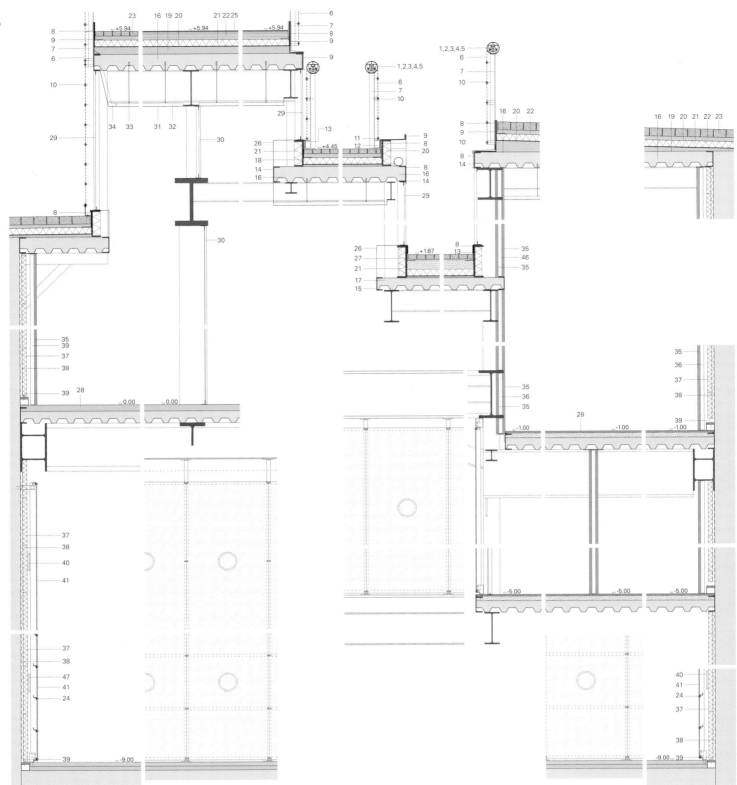

1. luminaria tipo encapsulite MT360 twin switched 21W longitud 965mm ø70mm double lámpara T5 trifósforo
2. pieza de sujección de nylon6. sujección con tornillos M5x20
3. plentina metálica compuesta por una 120.60.3, 60.30.3 soldadas
4. tubo metálico ø150.60.3mm(corte según las pendientes de las rampas)
5. pieza compuesta de cierre de seguridad del sistema de iluminación compuesto por dos plentinas de acero galvanizado de 130.30.3mm soldadas a una chapa de acero deployé RECA 50.22.3,3mm ancho 90cm, sistema de sujección a partir de tornillos de seguridad antirobo
6. chapa de acero deployé RECA 50.22.3,3mm ancho doblado ø150mm longitud de doblado 300mm
7. perfil en T60.3mm de longitud variable según el tipo de barandilla(especificaciones en cada detalle específico 300mm)
8. chapa de acero galvanizado 3mm
9. chapón de acero galvanizado espesor 8 o 10mm(ver detalles parciales)
10. arandela circula ø40mm fijada mediante tornillería a la pieza de anclaje(10), uniendo dos mallas entre sí(06) cada 25cm
11. perfil de acero galvanizado L40.40.3mm mecanizado para el anclaje de la malla de deployé(06)
12. tubo de acero galvanizado 60.100.3mm soldado al perfil en T de la barandilla(07)
13. perfil de acero galvanizado L100.100.8mm
14. perfil de acero UPN 200. conformación del borde del forjado
15. perfil de acero UPN 160. conformación del borde del forjado
16. base estructural. forjado. e=20cm
17. base estructural. forjado. e=16m
18. perfil en C 330mm
19. formación de pendiente. e máx=20cm/e mín=2cm
20. lámina impermeabilizante. doble en refuerzos.
21. aislamiento térmico. paneles de poliestireno extruído. e=8cm. URSA-XPS de superficie lisa, mecanizado lateral media madera
22. capa de arena. agarre de los adoquines. e=3cm
23. adoquines de granito. 10x10cm, e=8cm
24. tornillaería vista avellanada(def. ø)
25. pavimento de caucho reciclado
26. chapa de acero galvanizada doblada espesor=3mm
27. perfil de acero en C 400mm
28. pavimento de resina
29. doble vidrio KNT 169 saint-goblin o similar. securit 10mm/12mm/stadip 4+4. directo a perfilería de acero con silicona estructural de alta adhesividad
30. vidrio securit 10mm. directo a perfilería de acero con silicona de alta adhesividad[+herrajes de acero inox. para mecanismo de puertas]
31. chapa de acero galvanizado 0,5mm(anclaje a partir de remaches)
32. perfil omega de acero galvanizado 20.40.20.2mm
33. varilla de acero 3mm roscada, sujecciones a los perfiles mediante tornillería, al forjado con tacos de alta resistencia
34. plentina de acero galvanizado especial para los bordes
35. doble placa cartón -yeso de pladur o similar 13+13(tipo: ver plano acabados)
36. estructura metálica para recibido de placa tipo pladur o similar+cámara
37. panel de lana de roca e=50mm d=70kg/m2 con velo mineral negro: rockwool 231.652 o similar. adherido o atornillado a lámina drenante
38. lámina drenante PVC sika SP20 o similar, de nódulos de 20mm. con solapes, atornillada con pieza especial sellante a pantalla(absorción acústica+aislamiento)
39. canal de drenage perimetral de acero galvanizado 100x100mm. tapa de acero galvanizado perforado
40. fluorescente circular 22w ø220mm philips master TL5 o similar
41. plancha de acero galvanizado 2000x1000, e=3mm, perforado 64% ø9.5mm, lacado según color a elegir

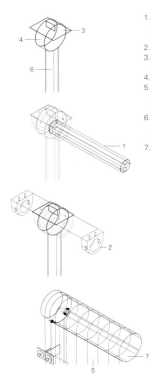

1. encapsulite lamp mt360 twin switched 21w. length 965mm ø70mm double lamp t5
2. fixing piece
3. composed metal piece 120, 160, 3 + 60, 30, 3 welded
4. metal tube ø150,3,60mm
5. curved deployed piece ø150mm and metallic pieces with double security screwing
6. deployed sheet RECA 50, 22, 3, 3mm width 100cm curved ø150mm curve length 300mm
7. steel T60 3mm variable length

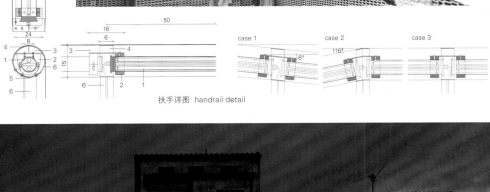

扶手详图  handrail detail

1. galvanized steel sheet 2000x1000, e=3mm, lacquered 2. galvanized steel sheet 2000x1000, e=3mm, perforated 64% ø9.5mm, lacquered 3. galvanized steel sheet 2000x1000, e=2mm, perforated 64% ø9.5mm, lacquered 4. T steel 45,80,4.5. Machining to screw(lacquered black) 5. Z galvanized steel 30,40,3(lacquered black) 6. C galvanized steel 30,40,3(lacquered black) 7. C galvanized steel 30,40,4(lacquered black) 8. countersunk screw (def, ø) 9. T steel 120,120,50,10. Fixed to structural base by screws 10. T steel 120,120,70,10. Fixed to structural base by screws 11. T of steel 220,120,50,10. Fixed to structural base by screws 12. rock wool panelling e=50mm, d=70kg/m²; Rockwool 231,652 or similar. Screwed on plasterboard 13. fluorescent light 22w ø220mm Philips Master TL5 or similar 14. double plasterboard or similar 13+13 15. metal structure for plasterboard or similar

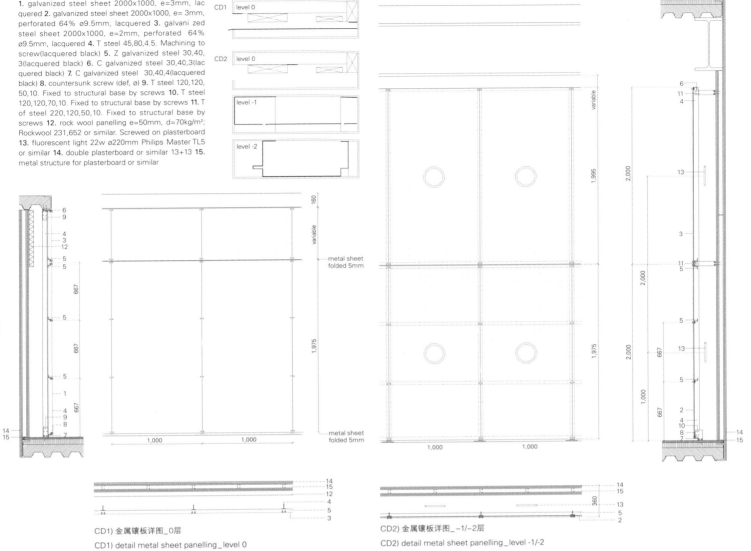

CD1) 金属镶板详图_0层
CD1) detail metal sheet panelling_level 0

CD2) 金属镶板详图_-1/-2层
CD2) detail metal sheet panelling_level -1/-2

项目名称：Teruel-Zilla  地点：Plaza Domingo Gascon, Teruel, Spain
建筑师：Mi5 Arquitectos, PKMN Architectures
结构工程师：Mecanismo Diseno, Calculo de Estructuras S.L. / 技术工程师：Maria del Carmen Nombela, Ana Macipe
设备工程师：Solventa Ingenieros Consultores S.L. / 岩土工程师：Geodeser S.A. / 地形学：Julia del Toro
施工单位：Fomento de Construcciones, Contratas / 甲方：Urban Teruel. Ayuntamiento de Teruel
表面积：urbanization_3,600m², built_3,005m² / 造价：EUR 7,574,459 / 竣工时间：2011.12
摄影师：©Miguel de Guzmán(courtesy of the architect)-p.152~153, 157 top, p.160~161, p.162~163
©Javier de Paz Garcia(courtesy of the architect)-p.150, p.154~155, p.157 bottom, p.159, p.161

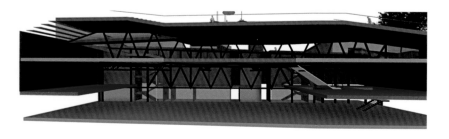

1. galvanized steel sheet 2000x1000, e=3mm, perforated 64% ø9.5mm, lacquered 2. galvanized steel sheet 2000x1000, e=2mm, perforated 64% ø9.5mm, lacquered 3. T steel 45,80,4.5. Machining to screw(lacquered black) 4. Z galvanized steel 30,40,3(lacquered black) 5. C galvanized steel 30,40,3(lacquered black) 6. countersunk screw (def, ø) 7. T steel 120,120,70,10. Fixed to structural base by screws 8. T of steel 170,120,50,10. Fixed to structural base by screws 9. rock wool panelling e=50mm, d=70kg/m²; Rockwool 231,652 or similar. Screwed on plasterboard 10. rock wool panelling e=50mm, d=70kg/m²; Rockwool 231,652 or similar. Screwed on Drain Sheet 11. fluorescent light 22w ø220mm Philips Master TL5 or similar 12. drain sheet PVC Ska SP20 or similar 13. reinforced concrete piling e=700mm

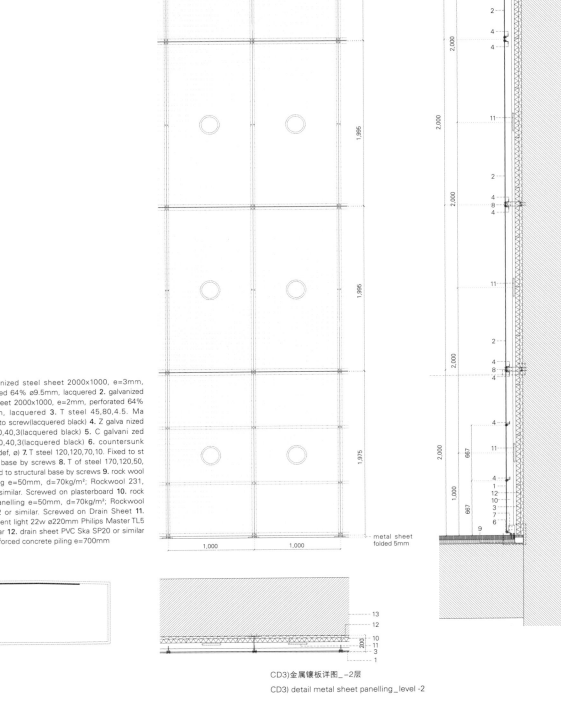

CD3) 金属镶板详图_-2层

CD3) detail metal sheet panelling_ level -2

## Georges-Emile-Lapalme文化中心

Menkès Shooner Dagenais LeTourneux Architectes + Provencher Roy+Associés Architectes

艺术广场是一个缩影，反映了过去四十年的思想和文化发展的历程，已成为了蒙特利尔公共场所的标志。文化中心的入口及大厅于1972年启用，由此可以进入多种演出场地和地下空间。进入文化中心相对简单，但它从未真正成功地突出其可提供的广泛的文化活动及服务，如演出大厅、售票处、时装店、饭店及四通八达的人行道。因为使用频率高，文化中心的入口随着岁月的流逝已斑驳不堪，蒙特利尔艺术广场目前正在寻求设计新的华盖、主入口以及通讯设施，以便建造出一个有独特风格、久负盛名的文化中心，用来举办广泛的文化活动。

新的设计以艺术广场现有的图形及建筑规模为基础，注重创造连贯的空间，与建筑的特点及大堂所在位置高度和谐。那些重新定义了文化范畴的智能多媒体显示器取代了电子广告及广告牌，显示了艺术家的才能和智慧。文化中心延续了艺术广场固有的特点和设计原理，强调社区、广泛的文化范畴、地下空间以及户外生活空间，进而重申了艺术广场的本质。大量的结构上的挑战、内部的整修以及都市设计风格共同促成了这座在原有建筑基础上修建的高质量的建筑。一系列的门槛、过渡点、土方工程、墙体和包含了照明系统、视听系统的天花板强化了空间层次感，创造出令人愉悦的的通道。这一空间被规划用于放映动画片和进行展览。地下空间、户外场所、天窗和展览中的艺术作品由一些开口连接起来，给人的整体感觉是一个巨大的、面向所有人开放的文化大厅。

如今，这一文化中心将给公众带来有助于发现和休闲的氛围。每天有三万五千人，每年有八百万的行人、消费者和赞助商从入口进入，经过大厅，走进文化中心。建筑要挑战的就是将魁北克最大的文化综合体转变为民众的生活环境。活动的多样性和选择的自由性已成为这一建筑的永恒特点：可及性是关键，而架起沟通的桥梁是它的使命。

### Georges-Emile-Lapalme Cultural Center

Place des Arts emerged as a model reflecting the ideological and cultural contexts of the last forty years and became an emblem of public spaces in Montréal. The entrance way and the foyer of this cultural complex first became functional in 1972, providing access to the various performance spaces as well as an underground network. The complex was easily accessible to users, but it never really succeeded in highlighting the wide range of cultural activities and services available, such as performance halls, box offices, boutiques, restaurants, and extensive pedestrian walkways. With such a high level of access and use, the complex suffered the wear and tear of time. The Place des Arts de Montréal was seeking new designs for the marquee, the main entrance, and the communication facilities in order to create a prestigious cultural center with a unique identity offering a wide range of cultural activities.

Based on the existing graphic and architectural dimensions of Place des Arts, the design focused on creating a coherent space in full harmony with the identity and the location at the lobby level. The eclectic advertising and billboards were replaced by judicious multimedia displays that redefine the cultural spectrum, testifying to the abilities and talents of our artists. The Espace culturel Georges-Émile-Lapalme provides continuity with the intrinsic features and design principles of the complex – the focus on the community, a broad cultural spectrum, underground and outdoor living space – thus reaffirming the identity of Place des Arts. The considerable structural challenges, the interior renovations and the urban design all helped produce a high quality architectural work, based on an existing building. Space is punctuated by a series of thresholds indicating points of transition, ground works, walls, and ceilings – that are illuminated by the lighting and au-

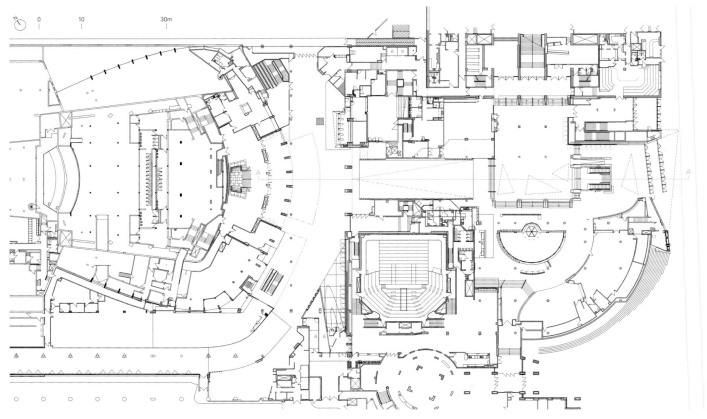

地下一层 first floor below ground

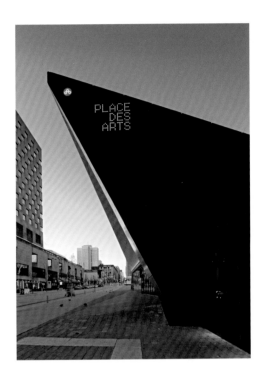

diovisual system, creating passage ways that are enticing and appealing to the senses. Spaces have been organized for animation and shows. There are openings that link underground spaces and the outdoors, as well as skylights and the works of the artists on display. The overall impression is an immense cultural foyer, available to all.

Today, the Espace cultural Georges-Émile-Lapalme introduces the public to an atmosphere that is conducive to discovery and leisure. The entrance way and foyer to the complex provide an access way, through which thirty-five thousand users transit on a daily basis, and eight million pedestrians, customers and patrons on an annual basis. The architectural challenge was to transform the largest cultural complex in Quebec into a living milieu. The diversity of its activities and the freedom to choose have become its most sustainable features: accessibility is the keyword, and the pathways of communication, its ambassadors.

Menkès Shooner Dagenais LeTourneux Architectes + Provencher Roy+Associés Architectes

项目名称：Georges-Emile-Lapalme Cultural Center
地点：Quebec, Canada
建筑师：Menkès Shooner Dagenais LeTourneux Architectes, Provencher Roy+Associés Architectes
项目主管：Jean-Pierre LeTourneux, Claude Bourbeau
艺术总监：Luc Plamondon
项目团队：Menkès Shooner Dagenais LeTourneux Architectes_Jean-Benoît Tremblay, Thomas Bernard Kenniff, Quinlan Osborne, Vincent Lauzon, Nils Rabota / Provencher Roy+Associés Architectes_Francis Berthiaume, Philippe Mizutani, Pascal Lessard, Véronique De Bellefeuille / Moureaux Hauspy + Associés Designers_Brent Swanson
照明工程师：François Roupinian
多媒体：Érick Villeneuve
标志设计师：Louis-Charles Lasnier
建筑面积：7,840m²
竣工时间：2011
摄影师：©Stéphane Groleau(courtesy of the architect) (except as noted)

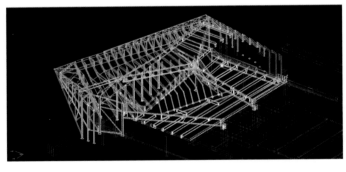

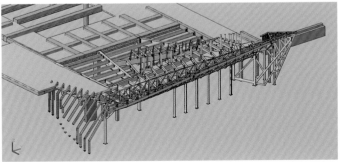

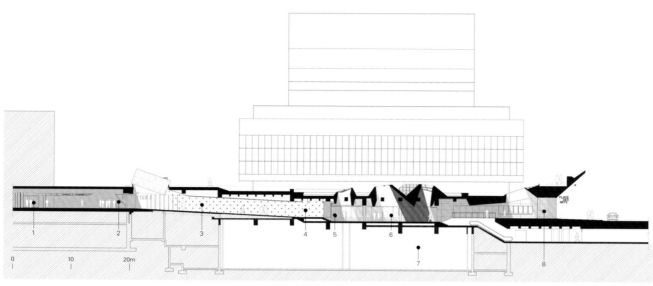

1 Wilfrid-Pelletier大厅入口 2 售票处 3 沉浸式多功能区域 4 展览区/节日接待处
5 平台/酒吧 6 入口大厅 7 排练大厅 8 通往圣凯瑟琳大街的主入口

1. entrance to salle Wilfrid-Pelletier 2. ticket office 3. immersive-multimedia zone 4. exhibition area/festival reception
5. terrace/wine bar 6. entrance hall 7. rehearsal 8. main entrance to Rue Ste-Catherine

A-A' 剖面图 section A-A'

# 南阿尔伯塔理工学院停车场

Bing Thom Architects

南阿尔伯塔理工学院新的三层停车场将分散的校园停车场整合成了一个中央设施，腾出了可开发的空地，恢复了校园生活的勃勃生机。

考虑到这个三层停车场的大小，首要的设计挑战就是要调和它的实际规模和预期规模。设计团队将此停车场置于一个现有的山坡上，从而使该停车场与周围景观有机地连接起来。只有东面和南面是可以完全看到的，形成了一个绿顶的四季运动场，扩展到整个结构。这一绿顶设计带有哥特式传统大厅的风格，从而确保了新停车场与建于1921年的大楼在风格上的一致性。玻璃锥体结构将楼梯连接起来，形成了天井，使自然光线可以进入停车场内部。由于充分考虑了"历史"这一校园灵魂要素，新停车场选址慎重，使得校园与卡尔加里市中心之间严密的景观与视觉联系得以保持。

奇特的立面延续了景观与设计之间的关系。东立面和南立面覆盖着半穿孔的金属幕墙，使得自然光线和通风能够进入停车场内部，与此同时，还形成了一个巨大的外层艺术品，与阳光形成互动，既描绘着阿尔伯塔草原的上空，又融入天空之中。在设计立面的过程中，设计团队找到了一种可以集艺术性和实用性为一体的解决办法，该设计使本可能巨大而毫无生气的停车场变得人性化，以一种简单而优雅的方式将一个普通的建筑类型转变成了一座与景观融为一体的建筑。

项目名称：SAIT Polytechnic Parkade
地点：SAIT Polytechnic main campus, Calgary, Alberta, Canada
建筑师：Bing Thom Architects
项目经理：MKT Arkle Development Management Inc.
幕墙艺术家：Rod Quin
结构工程师：Cohos Evamy Integrated Design
机械工程师：AECOM Canada Inc.
电气工程师：Crossey Engineering Ltd, Beaubien Maskell Engineering Inc
岩土工程师：EBA Engineering Consultants Ltd.
照明工程师：Crossey Engineering in association with William Lam
景观建筑师：SWA Group in association with IBC Landplan
施工单位：PCL Construction Management Inc.
甲方：SAIT Polytechnic
用途：parking, soccer field
建筑面积：35,396m²
造价：92 million USD
施工时间：2006.12—2009.9
摄影师：courtesy of the architect(except as noted)

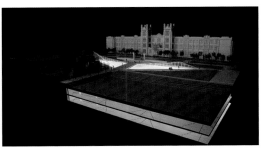

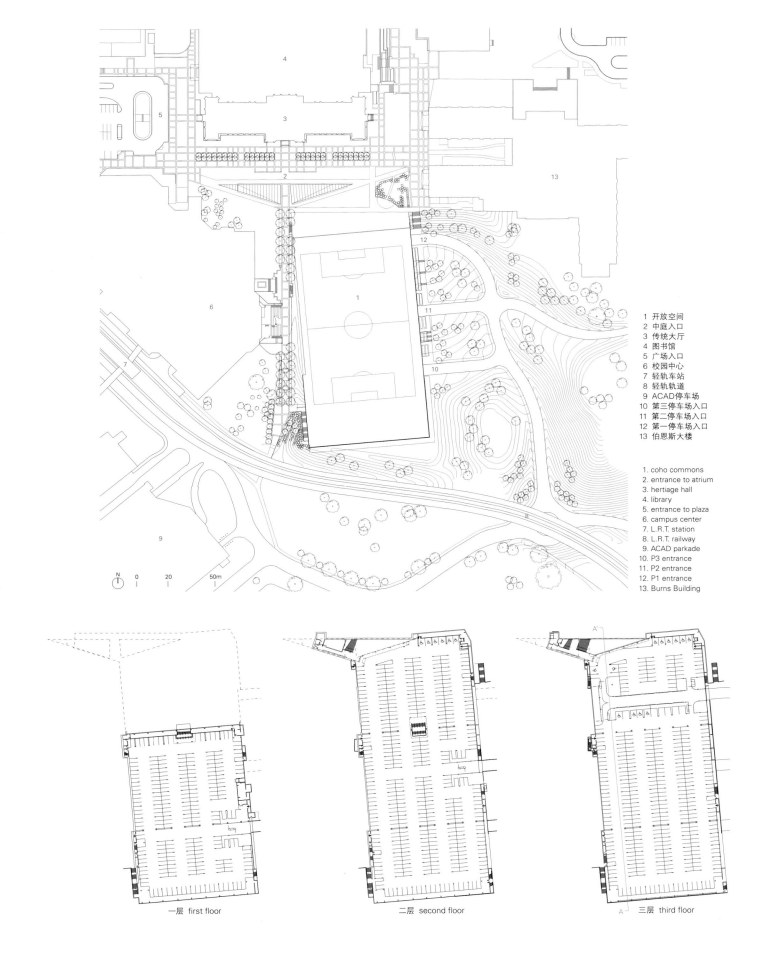

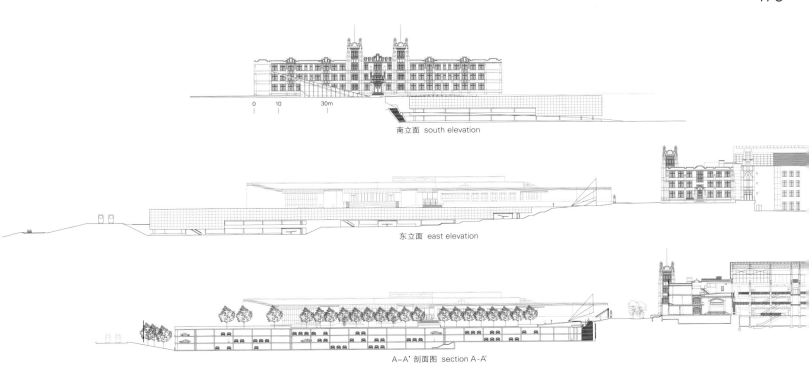

南立面 south elevation

东立面 east elevation

A-A' 剖面图 section A-A'

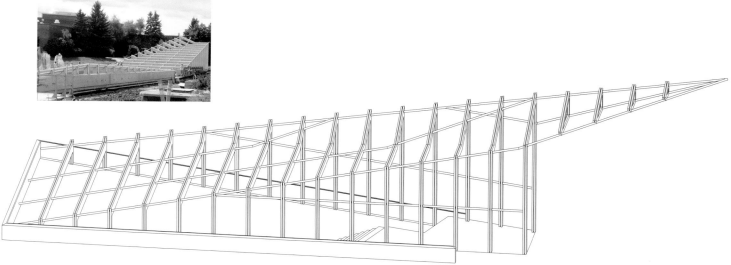

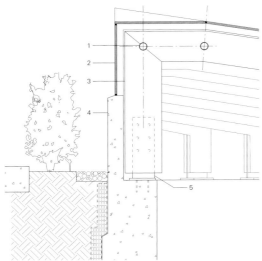

中庭西翼沿外墙及玻璃面剖面图
section through external wall and glazing, west wing of atrium

剖面图_玻璃谷 section_glass valley

1. painted steel pipe
2. single glazing
3. moment framed wood structure
4. concrete curb
5. knife plate welded to base plate
6. aluminum extrusion connector

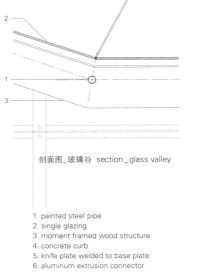

木结构剖面图_典型玻璃连接处
section @ timber structure _ typical glazing connection

## SAIT Polytechnic Parkade

SAIT Polytechnic's new three-level parkade consolidates the sprawled distribution of campus parking into a central facility, freeing up developable land and reinvigorating campus life.

Given the size of the building, the primary design challenge was mitigating its actual and perceived scale. The design team established a profound relationship between the building and surrounding landscape by nestling the building into an existing hillside. Only the east and south sides are fully visible, creating a green roof year-round sports field that extends over the structure. The green roof is at grade with the gothic Heritage Hall ensuring the parkade does not obstruct views of the historic 1921 building. Glass pyramids articulate the stairways and create atria that allow natural light down into the structure. By siting the building carefully, a critical landscape and visual relationship between the campus and downtown Calgary was preserved, respecting the historic heart of the campus.

Continuing the relationship between landscape and design is the pixilated facade. The east and south facades are clad in semi-perforated metal screens, allowing natural light and ventilation into the interior of the parkade, while simultaneously creating a giant exterior art piece that interacts with sunlight to both depict, and blend into, the Alberta prairie sky. In creating the facade, the team set out to find a design solution that combines art and utility. The design humanizes what might otherwise be just an enormous, soulless structure essentially turning a quotidian building type into something that becomes part of the landscape in a simple and elegant way. Bing Thom Architects

## >>68
**Estudio de Arquitectura Javier Terrados**
Javier Terrados was graduated from the ETSA(School of Architecture of Seville) and received a Master of Architectural Design from the graduate school of Cornell University. Coming back to Spain, have taught architectural design at the ETSA while running his own office in Seville. Has been a visiting teacher at several architecture schools in Spain and other countries. His work has been developed mainly in public architecture along with some singular houses. Has been awarded many prizes in Spain and some other European countries.

## >>46
**General Design**
Shin Ohori was born in Gifu, Japan in 1967. Studied architecture at the Musashino Art University in 1990 and received a master's degree from the same university in 1992. Established General Design in 1999.

## >>150

**Mi5 Arquitectos** right page top **+ PKMN Arquitectura** right page bottom

Manuel Collado Arpia[left] and Ignacio Martin Asunción[right] are founders of the Madrid-based architectural office Mi5 Arquitectos since 1999. Have won several competitions and their work has been published in media throughout the world. Have been taught at the Architectural Association School of Architecture in London at the summer school workshop 2011 and at the Architectural Polytechnic Universities of Alicante, in addition to having participated in several juries, lectures and exhibitions such as Venice Biennale, RIBA London, IVAM Valencia among others.

PKMN Arquitectura is an architecture office based in Madrid since 2006. The four partners: David Perez Garcia, Carmelo Rodriguez Cedillo, Enrique Espinosa Perez, and Rocio Pina Isla[from the left] do research into technology-typology-construction simultaneously. Love exploring new architectural fields connecting citizens, identity, pedagogy, communication, game, action and cities, specially throughout the strategies of participation, mediation and social innovation, and experimental active learning process.

## >>118
**Handel Architects**
Frank Fusaro received a Bachelor of Science
from the Ohio State University, and earne
Architecture from the University of Texas Scho
and Environmental Design. Prior to joining Ha
he spent 11 years at Buttrick, White & Buritis
he held dual responsibilities as project desig
architect on a variety of institutional and com
Joined Handel Architects in 1997 and became a
His projects have been published in Archit
Interior Design Magazine and The Wall Street J

## >>140
**nsamble Studio**
Was established in 2000 by Antón García-Abril<sup>left</sup> together with his partner Débora Mesa<sup>right</sup>.
ntón García-Abril has been an associate professor teaching architecture at Polytechnic
niversity of Madrid for a decade. Established his own studio with the aim of forwarding
eir view on urban development in 2000. Is currently developing a second doctoral thesis on
Stressed Mass" in civil engineering department of Polytechnic University of Barcelona. Are
so in the process of setting up a research laboratory at MIT and the POPLab (Prototypes of
refabrication Laboratory).

## >>34
**ouis Paillard Architect & Urbanist**
ouis Paillard was born in Paris, France in 1960. Graduated from the Ecole d'Architecture
aris-La-Villette in 1988 and became a certified architect. Established his own office in 2003.
elieves strongly in environmental approach, integrating the human social heart of the proj-
ct. His office integrates environmental quality from the design process, in order to anticipate
 minimize the environmental impact of architectural intentions initial technical choices, cost
 construction and long-term financial impact on the operation of building.

## >>60
**Jaques Moussafir Architectes**
Jacques Moussafir was born in 1957. Studied at the Ecole d'Architecture de Paris-Tolbiac and art history at Sorbonne University with Daniel Arasse. Established his firm in 1993 after training over a period of 10 years in the architectural office of Christian Hauvette, Henri Gaudin, Dominique Perrault and Francis Soler. Gave lectures at a number of European schools such as Polytechnic University of Barcelona and the school of architecture in Valle. Has been continuing lecturing as an associate lecturer at several universities in Paris since 2003.

## >>56
**Teatum+Teatum**
Was founded in 2011 by Tom and James Teatum. Produce innovative contemporary architecture, focused on exploring the experiential qualities and social opportunity of space. Their approach seeks a contemporary architecture that interacts with its physical and historical context, spatially engaging to users. Explores the possibility of each project through an intuitive and analytical study of the brief, the program and site to produce a specific architecture full of possibility. Both directors are teaching in London Universities. Tom is a unit tutor at Royal College of Art and is a steering group member of the RIBA Building Futures Think Tank.

## >>74, 126
**OFIS Arhitekti**
Is an architectural practice established in 1996 by Rok Oman [right] and Špela Videcnik [left].
Both graduated from the Ljubljana School of Architecture in 1998 and Architectural Association School of Architecture in London in 2000. Their project always starts with the search for a critical issue with the program, site or the client. Their design is not about surpassing, confronting, ignoring or disobeying the rules and limitations of each project. Rather, they plunge right in the middle of them and obey the law literally, word by word, if need be and at times even exaggerate it. In their work, restrictions become opportunities for an architectural system. In that sense, they become subversive, turning the limiting conditions into operational tools and so exposing all of the different possibilities.

## >>164
**Menkès Shooner Dagenais LeTourneux Architectes**
**Provencher Roy+Associés Architectes**
Jean-Pierre Letourneux is a 1983 graduate of the School of Architecture of the University of Laval. Founded the firm Dupuis LeTourneuz Architectes which today operates under the name of Menkès Shooner Dagenais LeTourneux Architectes. His capacity is to understand urban issues, his deep-seated knowledge of construction and his talent as a designer help to drive the agency.
Claude Bourbeau joined Provencher Roy+Associés Architectes in 2005 as a shareholder following the firm's acquisition of Beauchamp & Bourbeau. Acts as a project manager, designer and specialist in sustainable development, following an approach based on the LEED philosophy. Works at ground level on a wide range of construction, expansion and redevelopment projects involving office buildings, recreational and health and education facilities.

## >>82
**Alberto Campo Baeza**
Was born in 1946 in Spain. Graduated from ETSAM(Madrid Technical School of Architecture) in 1971. Has been teaching at several universities and institutes. Has ever completed a number of significant projects. His works have been published in major architectural magazines and exhibited in major cities. Most recent works are Cultural Center in Cobquecura, Chile and Olnick Spanu Museum in New York.

### Diego Terna
Received a degree in architecture from the Politecnico di Milano and has worked for Stefano Boeri and Italo Rota. Has been working as critic and collaborating with several international magazines and webzines as an editor of architecture sections. In 2012, he founded an architectural office, Quinzii Terna together with his partner Chiara Quinzii. Currently is an assistant professor of Politecnico di Milano and runs his personal blog L'architettura immaginata (diegoterna.wordpress.com).

### Simone Corda
Is an architect based in Sydney, Australia. Explores the themes of contemporary architecture through researches and projects at different scales and cross sectors. Referring to the architecture of Australia and New Zealand, he is currently focusing on the flexibility of housing as the key concept for sustainability. Part of his PhD thesis about Glenn Murcutt's work has been already published in the Italian magazine Area. Contributes to the Faculty of Architecture at the University of Cagliari enthusiastically with regular seminars and lectures at the Faculty of Architecture in Alghero, National Center of Research and Festarch event.

### Paula Melâneo
Is an architect based in Lisbon. Graduated from the Lisbon Technical University in 1999 and received a master of science in Multimedia-Hypermedia from the École Supérieure de Beaux-Arts de Paris in 2003. Besides the architecture practice, she focused on her professional activity in the editorial field, writing critics and articles specialized in architecture. Since 2001, she has been part of the editorial board of the portuguese magazine "arqa – Architecture and Art" and the editorial coordinator for the magazine since 2010. Has been a writer for several international magazines such as FRAME and AMC. Participated in the Architecture and Design Biennale EXD'11 as an editor, part of the Experimentadesign team.

## >>170
### Bing Thom Architects
Bing Thom is the principal of Bing Thom Architects, a Vancouver-based firm he founded in 1982. Is a graduate of the University of British Columbia and the University of California at Berkeley and a dedicated and artful city-builder whose global reputation has risen in consort with that of Vancouver. Believes in a holistic approach to architecture and attributes much of his success to his involvement in all aspects of building. This includes not only an understanding of the programmatic, technical and tectonic aspects of building, but also urban planning, development and construction. While international in outlook, his design approach is always closely rooted in a deep understanding of the culture and history of each project's community.

## >>104
### Kraus Schönberg Architects
Tobias Kraus[left] and Timm Schönberg[right] founded Kraus Schönberg Architects in London, UK and Constance, Germany in 2006. Have been visiting lecturers and critics in several universities. The Royal Institute of British Architects(RIBA) has named their project H27D a winner of the 2012 RIBA Awards. Two partners have been selected as the most promising and emerging European design talents for 2009 and were shortlisted for the British Young Architects of the Year Award 2008.

C3:Variation and Transition
All Rights Reserved. Authorized translation from the Korean-English language edition published by C3 Publishing Co., Seoul.

ⓒ 2013大连理工大学出版社
著作权合同登记06-2013年第16号

版权所有·侵权必究

图书在版编目(CIP)数据

在城市中转换 / 韩国C3出版公社编 ; 于风军等译. — 大连 : 大连理工大学出版社, 2013.3

书名原文: C3:Variation and Transition
ISBN 978-7-5611-7737-2

Ⅰ.①在… Ⅱ.①韩… ②于… Ⅲ.①建筑设计—作品集—韩国—现代 Ⅳ.①TU206

中国版本图书馆CIP数据核字(2013)第049665号

出版发行：大连理工大学出版社
　　　　　（地址：大连市软件园路80号　邮编：116023）
印　　刷：精一印刷（深圳）有限公司
幅面尺寸：225mm×300mm
印　　张：11.75
出版时间：2013年3月第1版
印刷时间：2013年3月第1次印刷
出 版 人：金英伟
统　　筹：房　磊
责任编辑：张昕焱
封面设计：王志峰
责任校对：张媛媛

书　　号：ISBN 978-7-5611-7737-2
定　　价：228.00元

发　行：0411-84708842
传　真：0411-84701466
E-mail: 12282980@qq.com
URL: http://www.dutp.cn